HORTUS CURIOUS

MICHAEL PERRY

AKA **MR. PLANT GEEK**

HORTUS CURIOUS

DISCOVER THE WORLD'S MOST WEIRD AND WONDERFUL PLANTS AND FUNGI

CONTENTS

STEP INSIDE MR. PLANT GEEK'S WORLD OF PLANTS

Plants continually excite and surprise me.

I've felt like that ever since the hazy days of childhood, when I first slipped a runner bean seed between some paper towel and the side of a jar and placed it on the windowsill in Mrs. Garden's classroom. Yes, indeed, my primary school teacher was called Mrs. Garden!

The first cutting I ever took was a tradescantia with silver stripes. I conscientiously potted it up and took it home, beaming with pride as I strutted into the family home. Days later, I decided to move this "houseplant" to my patch of garden, which was my first gardening mistake. The second was when I stood back to dig my hole and trampled on the poor thing. I cried for days.

Thankfully, though, this didn't dampen my spirits and, as a boy, I spent many happy weekends with my grandparents as they tended their large garden and preened dahlias to exhibit at local flower shows. Here, I learned my skills. Most importantly, I also learned about plants. We grew the Sensitive plant (*Mimosa pudica*), which had me gasping when the leaves curled up as I touched them. We also spent time rearranging the flowers of the Obedient plant (*Physostegia virginiana*), my eight-year-old self toying with nature's best-hinged blooms.

As I became a shy teenager, I carried on hanging out with my grandparents much of the time. Weekends were spent potting and pruning rather than playing video games or watching soccer and rugby. I wasn't a cool kid. I didn't have the right jacket or shoes and preferred to keep a low profile, finding my only refuge in plants. Fast-forward more years than I care to mention and plants magically became the next cool thing. I still find that hard to believe—although, of course, I'm very happy about it.

Aged 18, with a diploma in horticulture under my belt, I started working for a mail-order gardening company, where I was propelled through the ranks thanks to a love of plants and my ability to spot new ones. I became the guy behind the regeneration of the TomTato® (an internet search will return a photo of me with this amalgamation of tomato and potato). The Egg and Chips® plant (see page 114) came next, with eggplants and potatoes on one plant. But my job wasn't just to make plants fun—it was also to solve gardening problems for home gardeners. And, aside from being very marketable products, plants amazed me. They still do—the amount they achieve without the ability to move is quite exceptional. Animals can easily duck out of the way of predators, but plants don't (always) have that luxury.

My curiosity for plants knows no bounds and so, many years ago, I pulled together all my passions and began hosting "The Weird and Wacky Plant Show," a larger-than-life show about plants. Audiences gasped at the sight of the Coco de mer (see page 130), smirked at *Clitoria ternatea* (see page 164), and had their taste buds tickled by the Miracle berry (see page 126). From Canada to Japan, I sparked an interest in plants in the most plant-blind of people.

In this book, you'll meet many plants, each fascinating in its own way. In the **Superheroes** chapter, you'll marvel at the most momentous of them as we celebrate the largest, smallest, luckiest, and even the one that's the greatest dancer. We will explore how they became what they are, and why.

In the **Mistaken Identity** chapter, you'll encounter the plants that are often the subject of memes. They're the lookalikes, those that resemble flying ducks, or penguins running over a mountain, or even a dead man's hand reaching out from the forest floor. Many of them have had to become experts at disguise, but not always for the reason you might expect. With limited resources at their disposal, plants have to do all they can to attract pollinators. It's like they're shouting "hey you, come get me." They put it all out there.

But there's more: in the **Plants Behaving Badly** chapter, you'll find out about insect-trapping plants that operate underwater, plants that are flammable, and others that smack unsuspecting bees with pollen. It's a crazy world. Of course, sex is often involved, and so as you delve into the **X-Rated Plants** chapter, you'll need to put your inhibitions to one side. Prepare yourself for general indecency, squirting cucumbers, and devil's fingers.

There's a more serious side to the book, too. In the **Greater Good** chapter, we look at the most commercially successful plants the world has ever seen. Whether they're used as materials or as foodstuffs, it's a fascinating collection of plants with myriad uses, from latex to lubricant, Filipino formal wear to fancy skincare.

You see, I have no hobby other than plants: I am that two-dimensional. Yes, I like walking and traveling and eating, but my partner knows that every vacation destination I choose has an ulterior botanical motive. The main thing that inspires me is the

plants I'll see en route and those I'll get to devour. Plants never fail to beguile me. With their intricate relationships, and techniques for deception and extreme survival, they've got life all sewn up.

I am super lucky to have met a lot of plants, but there are still many that remain elusive. In some ways, you could say that this book is a bucket-list of plants that I want to spend time with. It's something you can share with me, because a fair few of them can be grown in your own backyard, so you, too, can marvel at nature up close and personal. You'll have your own personal ringside seat to the plant circus.

As I write this, I'm in a room that's an awkward hybrid of an office and a greenhouse, with trailing plants dangling from the curtain rod and a desk littered with diddy pots of succulents. I feel amazingly content. But, dear reader, you must know that I am no botanist, and this book was a learning curve for me. So I shall forever be thankful to a bevy of botanical friends who, from across a mix of time zones, helped me translate scientific speak into something I—and you— could easily digest. Without their assistance, I couldn't have created such a coherent yet fun book. I hope my analogies make sense to you and that they also make you giggle—from the dance floor of the common stinkhorn (see page 172), to the manipulative dating behavior of the flying duck orchid (see page 54).

My eight-year-old self would be marveling at how loved plants have became, and it's our duty to carry on protecting them at any cost. The diversity of our planet is key to our survival. The good news is that you can play your part at home by filling your life with color, and your outdoor space with pollinator-pleasing flowers, so that plants can continue their vital relationships with the creatures that visit them, live among them, and feed on them. For without plants, we are nothing.

PLANTS
BEHAVING
BADLY

01

Turn the page if you dare. Here are some of the most dangerous plants on the planet, causing harm to tiny insects, inquisitive mammals, and even fully grown humans. But as bad as their behavior may sometimes seem, it may be that in a big bad world that's exactly the way they need to be. And I am so here for it!

COMMON NAME ———————————

Waterwheel plant

HOW BIG IS IT?
About 16 in (40 cm) in length.

WHERE'S IT FROM?
Found in small populations in 50 spots across Europe, Africa, Asia, and Australia.

WHAT'S ITS NATURAL HABITAT?
Calm, shallow waterways.

HOW DOES IT REPRODUCE?
Pollination is infrequent, as flowering is unpredictable. Plants multiply by side-branching; when branches break off, they form new colonies.

CAN I GROW IT AT HOME?
Yes, you can, in a pond in summer.

If you thought insects are only at risk of being trapped and eaten by plants on land, you'd be mistaken—the water isn't safe either, especially not with this carnivorous plant. It catches and digests small aquatic invertebrates, such as mosquito larvae and water fleas, with traps like those of fellow carnivore, the Venus fly trap (see page 36).

The waterwheel plant's traps are just half an inch (1 centimeter) in diameter and are arranged in a wheel formation of usually 5–7 traps, around a stem that can reach about 16 in (40 cm) in length. A

THE FASTEST AQUATIC FLY TRAP

Aldrovanda vesiculosa

——————————————— LATIN NAME

The trap snaps shut in 10–20 milliseconds, making *Aldrovanda vesiculosa* the fastest mover in the plant kingdom.

barrier of bristles stops random debris getting caught up in each one (the plant would much rather have a live feast). When the unsuspecting invertebrate gets too close, trigger hairs are alerted and the trap snaps shut in 10–20 milliseconds, making *Aldrovanda vesiculosa* the fastest mover in the plant kingdom.

Blooming stems push the flower above the waterline in the hope of it getting pollinated.

Each trap has up to 40 trigger hairs, so even the most erratic swimmer gets caught. Once inside, the prey is hungrily broken down by enzymes. It can take a few days for the plant to digest its dinner.

Escape from the cold

The waterwheel plant grows entirely under water and is rootless, with no anchors to keep it in place. Its stems are hollow to help it stay afloat, but when cooler weather nears, the plant has a strategy to survive freezing conditions.

As its tender stems start to die off and decay, the growth tips start producing shorter, noncarnivorous leaves that form a tight bud (a turion) full of starch and sugars. They become heavy and, as they detach from the decaying stems of the plant, release their flotation gases and air, so they can sink to the bottom, where its warmer and they can hibernate.

When water temperatures rise above 54°F (12°C), the turions reduce their density and float toward the surface once more, where they resume growth and produce whorl upon whorl of hungry traps. Oh, the perils of working underwater—but *Aldrovanda* seems to have it sorted.

The waterwheel plant is the only species remaining in the *Aldrovanda* genus, which is a part of the well-known Venus fly trap family.

Whorl-like structures grow from the plant's stem and these are what give the plant its common name "waterwheel."

COMMON NAMES

Gympie-gympie, stinging tree

THE MOST PAINFUL PLANT

HOW BIG IS IT?
10–33 ft (3–10 m) tall, with a spread of about 3 ft (1 m).

WHERE'S IT FROM?
Australia, from Cape York to New South Wales, and Malaysia.

WHAT'S ITS NATURAL HABITAT?
Rainforests, especially along water courses and in gaps and clearings.

HOW'S IT POLLINATED?
Likely by bees.

CAN I GROW IT AT HOME?
No, and you wouldn't want to.

Planning a trip to Australia? Scared of spiders? Snakes? Well, they're the least of your worries when there are plants that can hurt you. Perhaps even kill you. Covered from head to toe in trichomes (hairs) containing a powerful neurotoxin, the gympie-gympie is probably the most venomous being in Australia. Its name means "stinging tree" in the language of the Indigenous Gubbi Gubbi people of South East Queensland, and it can sting you and inflict a burn, leaving pain that can persist for months. Even the dead leaves and decades-old dry samples can harm you. The trees can

Dendrocnide moroides
LATIN NAME

The bright fruits resemble mulberries, and are edible yet have never become quite as popular—I do wonder why.

also casually shed their hairs into the wild. Suspended in the air, they float around, poised for their next victim to inhale ....

The gympie-gympie's brittle hairs are loaded with toxins and are sharp and hollow, like a hypodermic needle. When the hair touches your skin, the tip fractures, injecting the venom. The skin becomes red and swollen, breathing can be difficult, and the pain is excruciating, intensifying for a few hours, and sometimes not easing for days and even months. Sleep is no longer an option. One victim said: "It's like being stung by 30 wasps." Another reported that "there's nothing to rival the pain— it's 10 times worse than anything else." And the bad news is that there's still no known antidote to the sting of the gympie-gympie.

Quick action with sticky tape (wax hair-removal strips are what the Queensland Ambulance Service recommends) can sometimes remove the hairs when they first pierce the skin. But if they stay there, they drip-feed their toxins whenever the affected area is triggered by rubbing or itching, and this can last for up to six months.

Edible mulberry-like fruit

Looking at the fruit, you'll notice that they resemble mulberries and, in fact, they're from the same family and are edible (the name *Dendrocnide* comes from the Greek for "tree" and "nettle" and *moroides* means "resembling a mulberry").

But who would dare eat them? The tree goes to extreme lengths to protect itself from predators—those stinging hairs are on the fruit, stem, branch, petiole, and leaf.

Some brave birds do snack on the fruit, though, and merrily distribute the tree's evil legacy via the seeds in their droppings. Many insects also feed on the leaves, and the gympie-gympie is a host plant for white nymph butterfly larvae. Some mammals, too, seem to be able to eat them without any visible side effects.

Unluckily for us humans, the gympie-gympie is quite a resilient plant, and an early colonizer of rainforest gaps. It can come to maturity and fruit at just 10 ft (3 m) tall, meaning peril is within easy reach. In fact, it often takes on the habit of a low-slung shrub, and these small plants have obtained the name "ankle biters." Unwary travelers brush by them, continuing on their journey with a bevy of toxic hairs coating their pant legs, backpack, or worse.

But it isn't all doom and gloom and I'm not really suggesting you avoid an Australian vacation. Only one human death from *Dendrocnide* has ever been reported, and that happened in New Guinea in 1922.

My brush with the plant

I met the plant once—at a safe distance behind glass—at the Hortus Botanicus Leiden, in the Netherlands. It was no more than about 10 in (25 cm) tall, and laden with scrumptious-looking berries covered in the ever so innocent furry hairs. I wouldn't like being on watering duty with the most dangerous plant in the world in residence. That's the best case for automatic irrigation I've ever heard.

> The pain is excruciating, intensifying for a few hours, and sometimes not easing for days.

The gympie-gympie's toxin is so unique in the plant world that scientists have named it after the plant— gympietides. The toxin is much like one found in some spiders and snails.

COMMON NAMES

Burning bush, gas plant

THE PYROMANIAC PLANT

HOW BIG IS IT?
Up to 16 in (40 cm) tall with a spread of 3 ft (1 m).

WHERE'S IT FROM?
Southern Europe and parts of Asia.

WHAT'S ITS NATURAL HABITAT?
Warm, open woodland.

HOW'S IT POLLINATED?
Primarily by bees, especially medium-to-large ones, *Habropoda tarsata* being spotted the most often.

CAN I GROW IT AT HOME?
Yes, in full sun or dappled shade and in good, moist, efficiently draining soil. But be warned—it takes a while to establish.

Now, I'm not encouraging you to be a botanical arsonist, but if that were your thing, then this is the plant to do it with. The burning bush is a quite phenomenal, yet little-understood, plant. It shimmers with volatile oils, which can ignite for a split second on a warm, still day. The effect is similar to a Bunsen burner lighting up momentarily during science class, albeit the flames are surrounding a most gorgeous flower spike. Thankfully, this Bunsen burner isn't accompanied by the teacher yelling for you to stop.

A native of temperate Eurasia, occurring from Spain to China, it makes a spectacular (and not at all dangerous) border plant.

Dictamnus albus var. *purpureus*

LATIN NAME

A botanical mystery

The bush's flammability was first documented by Swedish botanist Carl Linnaeus in 1753, when his daughter is said to have struck a match beneath the plant and it momentarily caught fire.

The plant is at its most flammable when the flowers fade and the seed pods begin to form. Its self-pyromania causes it no harm, since *Dictamnus* itself is rather flame retardant, and the ignition happens only momentarily.

But, considering how curious a phenomenon it is, the burning question of why the plant has developed these ethereal flammable fumes remains curiously unexplored by botanists.

One fanciful theory is that, by bursting into flames, it can burn off nearby competition—quite literally! By creating the spark on hot summer days, this bad boy has the potential to eliminate rival plants in the surrounding area by starting its own small-scale bushfire.

Other thoughts are that the flame is simply a side-effect of the oil production within the plant, or that it could help protect it against heat stress.

The question cannot, as yet, be extinguished. If, like me, you're a YouTube junkie, go check out the range of ignition videos online, where you can see this plant doing its party trick with your very own eyes.

A burning bush at home

Besides being highly flammable, the strongly-scented oils are also likely to have evolved to dissuade predators, including mosquitoes. So, you may want to pot one up for your summer patio—just make sure the container is really deep to accommodate the huge tap root, just like a carrot. If there's

HOW CURIOUS

Containing more than 100 noted chemical constituents, including those volatile oils, the humble *Dictamnus* is the ideal science project.

one thing the burning bush doesn't like, it's being disturbed, so avoid that at all costs.

But don't let that dissuade you from adding it to your gardening mood board. If that glossy, pinnate foliage (pinnate meaning paired leaflets coming from a single stem) doesn't seduce you, then the intricately detailed flowers will. Available in either white or purple-pink, the purple variety, in particular, has been selected for garden use and even landed the much-coveted Royal Horticultural Society's Award of Garden Merit. Way to go! Both varieties have the potential to light up your garden ... err, literally.

However, they are seldom seen in nurseries and garden centers because the production times are much longer than many other herbaceous perennials. And they need huge, inconvenient pots to accommodate that awkwardly-large tap root. In fact, I have never seen any for sale and I am a stalwart when it comes to plant sales. Perhaps they never got health and safety clearance.

If you can, it's worth hunting a few plants down, though: they are unfazed by pests and disease, and strong enough to be self-supporting. But make sure you label the plant clearly, just in case it spontaneously combusts one day. You never know ...

This bad boy has the potential to eliminate rival plants in the surrounding area by starting its own small-scale bushfire.

Narrow-lid pitcher plant

HOW BIG IS IT?
Pitchers can be 4 in (10 cm) in height and vines can reach 49 ft (15 m).

WHERE'S IT FROM?
From southern Thailand to New Guinea. This is one of the most widespread *Nepenthes* species.

WHAT'S ITS NATURAL HABITAT?
Damp forests, from sea level up to 7,000 ft (2,100 m).

HOW'S IT POLLINATED?
By small flies and tiger moths.

CAN I GROW IT AT HOME? Yes. Ideally, in a giant terrarium or with 75 percent humidity and constant warmth (sounds like a full-time job).

Here's a plant that at first glance seems carnivorous, but in fact is so docile that small insects and frogs regularly set up home inside its curious green rotundas. Unlike its insect-chomping relatives, *Nepenthes ampullaria* doesn't have any insect-luring nectar glands. Its pitchers are also different from the narrow type other *Nepenthes* species have. They have no lid to cast an ominous shadow over the inside and they're shaped more like funnels with a wide rim, having evolved to collect falling leaf litter. Drum roll please, *Nepenthes ampullaria* is in fact a vegetarian, and so perhaps is not as badly behaved as it first appears.

THE VEGETARIAN CARNIVORE

Nepenthes ampullaria

LATIN NAME

To maximize its leaf-collecting potential, it produces a massive number of pitchers, bunched up together on networks of tangled vines.

Pitcher this

So, once the pitcher has collected the leaf litter, how easy is it to break down? Not very. Fortunately, the pitcher plays host to a group of organisms, a kind of commune that includes, among others, mosquito larvae. This bunch of rogues might be considered to be exploiting the plant by moving in, except that they set to work chomping away, breaking down the leaf litter into nitrogen-rich waste, which the pitcher plant then greedily absorbs. The technical term for a plant that behaves in this way is a detritivore, meaning it

Long flower stems are common in carnivorous plants, allowing pollinators access to them far from the usually deadly pitchers.

The pitchers have been known to make a pretty nifty and compostable rice bowl, ideal for serving sticky rice dishes.

consumes detritus, which could be decomposing plants or, in some cases, even animal feces.

In its hunger for food, *Nepenthes ampullaria* leaves no stone unturned. To maximize its leaf-collecting potential, it produces a massive number of pitchers, bunched up together on networks of tangled vines, creating an often dense carpet of them across the forest floor.

Nepenthes ampullaria has the most diverse coloration of all the pitcher plants, appearing in every shade from green to red, as well as green with red spots. The plant clearly has a casual chameleon attitude, since there are no explanations for this variation in pitcher color or pattern.

A home to many

As well as mosquito larvae, other creatures make their home in the pitchers' green-carpeted lounge. In all, more than 60 unique organisms have been found to have either a mutualistic or a scavenging association with this *Nepenthes* species.

Crab spiders frequently rent some floorspace, as do micro frogs—one species of which, *Microhyla nepenthicola*, attaches its eggs inside and raises its tadpoles in the pitcher's watery lake below.

Like many *Nepenthes*, this species cross-pollinates quite readily with other species (the flirts!), giving rise to interesting hybrids. The flowers are mostly pollinated by flies, yet are seldom photographed, since the pitchers get *all* the attention.

Like many carnivorous plants, the flowers are held well above the "business area" of the plant, ensuring that visiting pollinators avoid the traps; although, as we know, they would actually be quite safe with *Nepenthes ampullaria*—these are merely a relic showing their evolutionary journey.

This has to be the coolest plant I know. If I spotted one in the wild, I would most likely scream in delight. Then I'd drop to my knees and really check out what's going on in the green room.

THE GREAT POLLEN CATAPULT

HOW BIG IS IT?
12–16 in (30–40 cm) tall, with a spread of 20 in (50 cm).

WHERE'S IT FROM?
Central and East Australia.

WHAT'S ITS NATURAL HABITAT?
Dry forests with poor levels of nutrients.

HOW'S IT POLLINATED?
By small, solitary bees and nectar-feeding bee-flies.

CAN I GROW IT AT HOME?
Actually, yes. It's said to be hardy to 14°F (–10°C) and can be grown easily as a border plant.

The trigger plant is so damn clever, leaving nothing to chance. When an insect lands on the flower, there's a gentle change in pressure. This causes a trigger from behind the flower to ping the "floral column," which is kind of like a magic wand that holds all the plant's reproductive parts. It douses the insect in pollen, leaving it dazed but unharmed. This all happens within 15 milliseconds and is quicker when the air is warm, rather than cold. The pollen-doused insect then flies off to another bloom, transferring pollen as it goes. Meanwhile, the trigger resets, ready to douse the next insect.

Stylidium graminifolium

LATIN NAME

~~~~~~~~~~

Wouldn't it be a joy to sit back and watch nature at work, as unsuspecting European bees are thrown around by this bright pink boss plant?

~~~~~~~~~~

The indigestible truth

For many years, the trigger plant was thought by some botanists to be protocarnivorous. Plants with this skill can capture, but not digest, all sorts of tiny insects that are much smaller than the creatures responsible for pollination. A thick, gluey substance called mucilage—composed of sugar polymers and water—on the plant's trichomes (hairs) can attract the insects and then suffocate them. They are simply stuck, with no option for escaping or feeding. Protocarnivory can also be a defense mechanism to stop parts of the plant being eaten. I'm hoping that a pot of *Stylidium* on my outdoor table would stop insects attracted to my yard in their flying tracks. It's definitely worth a go.

Grow your own

Stylidium graminifolium is easy to grow and makes a rather handsome border plant—and not just from a distance. Up close, you can enjoy the most gorgeous pink, butterfly-shaped blooms, which look pretty as cut flowers in bouquets. Never pick flowers from the wild, however, not least because they could arrive in your vase looking like an epitaph, with masses of unlucky little insects clasping to them.

A variety named "Little Saphire", with blue-green foliage, is listed by the Royal Horticultural Society. Try raising the plant from seed, which is quite readily available.

Once you have gotten your hands on some *Stylidium*, the plants are really undemanding, preferring low-nutrient soils, and often enduring bushfires in the wild. Why not make your planting part of a border circus-show, with the highly combustible *Dictamnus* (see page 20). They hate being disturbed, though, much like me during *EastEnders*. But, wouldn't it be a joy to sit back and watch nature at work, as unsuspecting European bees are thrown around by this bright pink boss plant? Now that's a nature documentary I'd abandon *EastEnders* to tune into.

Bushfires aid the germination of seeds in the wild, so if you're growing at home you may need to apply some smoke treatment.

Viewed from afar, these bright and eye-catching perennials make a stunning border plant in a garden.

COMMON NAME

Bullhorn acacia

HOW BIG IS IT?
33 ft (10 m) tall, with a spread
of up to 16 ft (5 m).

WHERE'S IT FROM?
Central America, particularly Yucatán
in Mexico.

WHAT'S ITS NATURAL HABITAT?
Wet lowlands, swamps, and riverbanks.

HOW'S IT POLLINATED?
By insects, although it's still not known
how they get away without being
attacked by the acacia ants.

CAN I GROW IT AT HOME?
Nope. And I'm not sure you'd want
to—it's pretty ugly.

**With a tangled network of terrifying
thorns, the bullhorn acacia offers
predators little chance of getting close.**
But it isn't just the plant that's behaving
badly. There are ants living inside those
thorns, and the plant has them
"brainwashed" into swarming with intent
whenever a predator gets too near. The
partnership between the bullhorn acacia
and these acacia ants (*Pseudomyrmex
ferruginea*) is, for me, one of the most
fascinating in the natural world.

The acacia's special relationship with
ants was observed on plants in Nicaragua
in the 1800s by English naturalist Thomas

A THORNY
RELATIONSHIP

Vachellia cornigera

LATIN NAME

Belt. Then, in 1973, American ecologist Daniel Hunt Janzen carried out field studies in Mexico. He soon concluded that the relationship this majestic tree has with acacia ants is actually vital to them both.

Give and take

Looking at the bullhorn acacia, the most notable thing about its appearance is its thorns, which do indeed resemble bull's horns. Acacia ants cut small holes in the thorns—which are hollow—and lay their eggs inside, so they can raise their larvae there once they hatch.

But the ants don't get to live rent free. The tree relies on their aggressive behavior to ward off predators, whether that's an animal or an insect: a goat might get stung on the head, a cricket crushed.

As well as attacking the predators, the ants also release a pheromone that scares them off. In tests carried out by Janzen with deer and cows, he observed that both avoided the trees for this very reason.

The plant thanks its ant army with a feast. More than half its leaflets have a detachable tip known as a "Beltian body," named after Thomas Belt, an English naturalist. Rich in lipids, sugars, and protein, these have one function—to feed the ants, whose larvae even have little side pockets on their bodies, into which the ants push them.

But the plant has an even more devious trick up its sleeve to secure the ants' loyalty.

> The partnership between the bullhorn acacia and these acacia ants is, for me, one of the most fascinating in the natural world.

HOW CURIOUS

Bullhorn acacia thorns are often strung into necklaces and belts in Central America.

Glands called extra-floral nectaries on its leaf-stalks ooze a carbohydrate-rich nectar. Alkaloids in the nectar include caffeine and nicotine, which the ants become addicted to. The plants even regulate the release of the substances to control the ants' behavior. But the ants couldn't be happier with their bed and board. They just need to swarm and look angry from time to time.

Self-imposed seclusion

Janzen also noticed the lack of other plants in the areas surrounding the bullhorn acacias. His research revealed that the ants clear away invasive seedlings from the base of the trees, before secondary plants have a chance to compete. They also snip away at neighboring branches that dare to grow into the acacias' airspace. And woe betide any ant army that tries to set up home on another's patch. That would almost certainly start a conflict!

Detachable yellow Beltian bodies feed the ants, while the hollow thorns add to the bed-and-board experience.

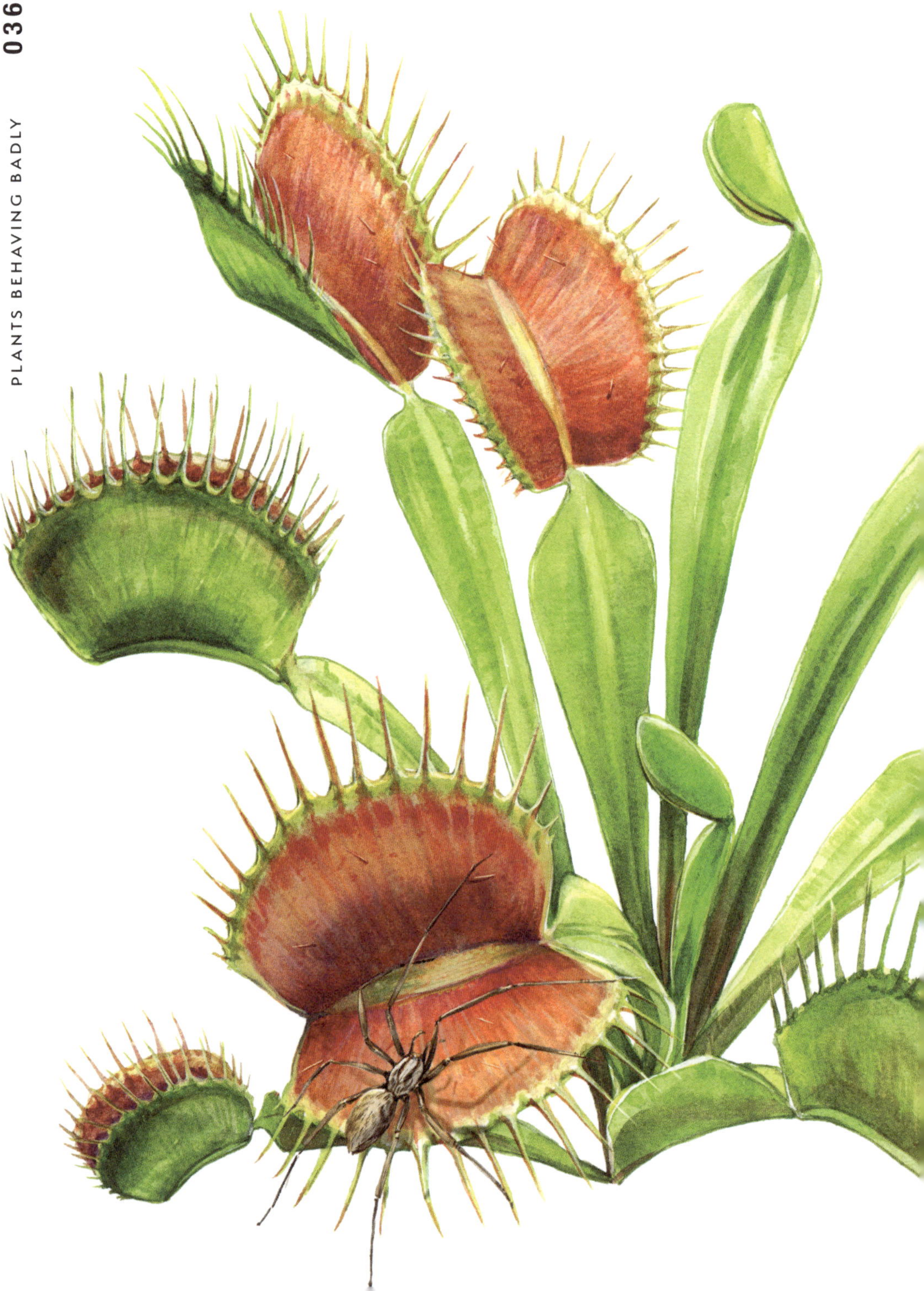

Venus fly trap

A PLANT INTENT ON MURDER

HOW BIG IS IT? 2–4 in (5–10 cm) tall, with a spread of 4–8 in (10–20 cm).

WHERE'S IT FROM? A strip of coastal land in North and South Carolina.

WHAT'S ITS NATURAL HABITAT? "Pocosin," a type of coastal savanna wetland.

HOW'S IT POLLINATED? By flying insects, such as longhorn beetles and sweat bees. Let's hope they don't linger and get caught.

CAN I GROW IT AT HOME? Yes, it's super fun and deceptively easy from seed. The plant comes to maturity within four years.

Looking at a Venus fly trap might well be the first time you realize how amazing plants really are. I bet you a fiver that you've tried to shove your fingers into one, too, masochistically hoping it would bite them off. Thing is, it won't. Despite its famously deranged and bloodthirsty alter ego in the 1986 movie *Little Shop of Horrors*, the Venus fly trap is picky about its food and doesn't crave blood all that often.

Its habitat in the subtropical wetlands of the east coast of the United States might sound luxurious enough, but the acidic soil there is devoid of most nutrients. So, over the centuries, the Venus fly trap has

Dionaea muscipula

The rows of spines clasp together, creating a cage-like trap that prevents a tasty meal escaping.

pair of lobes, with a hinge between them, poised to snap shut so the plant can eat.

The bright red insides shine like a beacon, and the trap emits a scent, though undetectable to the human nose. A matrix of tiny hairs on the lobes acts like a sensor to alert the plant to potential prey, but it's only when the hairs are triggered twice in around 20 seconds that the trap snaps shut. This prevents the plant from trying to crunch on lifeless forest debris or raindrops.

supplemented its diet with a wide menu of prey that includes flies, beetles, grasshoppers, spiders, worms, and the odd small frog.

The murder plant, as I like to call it, has numerous subterranean stems supporting a rather stupendous leaf blade. This leaf is part photosynthesizing petiole and part

The death trap

So, picture the scene: it's a calm day in the Carolinas and a leggy spider strolls innocently across the plant, stopping to rest on the surface of the pad. If the spider should then stumble, triggering those hairs, the pads go from convex to concave in the blink of an eye (technically, in one-tenth of a second). The rows of spines clasp together, creating a cage-like trap that prevents a tasty meal escaping, unless the prey is small and agile. In which case, it probably wouldn't be a wholesome enough meal for the plant, so it doesn't matter if it wriggles free.

I bet you hadn't thought about a Venus fly trap's flowers, but standing tall above the deadly traps are white flowers for insects to come and pollinate.

The speed the trap snaps shut varies from plant to plant and is faster, for example, on warm days. It can also be an indicator of plant health.

If an insect nudges the trigger hairs twice in around 20 seconds, the trap snaps shut around its prey.

Naturally, our sacrificial spider will attempt to break loose. Once it has triggered the hairs another five times, the trap knows it has a substantial meal in store. The lobes then force together to form a stomach of sorts, with the digestive glands firing up to break down our "itsy bitsy." This eating process takes about 10 days as the glands secrete an enzyme to decompose the prey and soak up the nutrients. After this, the trap reopens and awaits the next victim. Plants can live up to 30 years, so they can fit in quite a few gourmet meals within that time.

Protecting the species

Sadly, the US Fish & Wildlife Service has had to place the Venus fly trap under an Endangered Species Act. Plants are endangered by poaching (although, thankfully, most sold these days are raised by micropropagation or seed). Changes in land use have also led to habitat loss. Previously, Venus fly traps depended on naturally occurring forest fires to remove competing plants, since they themselves are quite flame retardant, reemerging from the fleshy leaf bases beneath the soil. With fire-suppression laws in force, they now don't have the chance to prosper.

There are many different types of Venus fly trap, some with flat traps, some with bigger teeth, and some which are almost black. Look out for 'B52' and 'South west giant', which have the biggest (and scariest) traps. The Venus fly trap is one freaking cool plant (I bet I get told off for swearing).

Dynamite tree, monkey no climb tree, sandbox tree

THE NOISIEST PLANT

HOW BIG IS IT?
197 ft (60 m) tall while the canopy can reach 33 ft (10 m) or more in width.

WHERE'S IT FROM?
Originally from South America, and found across the Amazon, it has also been introduced elsewhere.

WHAT'S ITS NATURAL HABITAT?
At the edge of the forest and in gaps; anywhere it can find a little space.

HOW'S IT POLLINATED?
Bats take care of that.

CAN I GROW IT AT HOME?
No way, José.

You thought all trees were quiet and just kept to themselves, didn't you? The most noise they might make is a rustling of leaves every now and then. Not so. *Hura crepitans* produces seed pods that violently and noisily crack open, exploding forth at 160 mph (260 kph), and ricocheting off trees in the forest, hitting trunks (and, hopefully, not your head) with a force and noise akin to a gunshot. No wonder it's sometimes called the dynamite tree.

The pumpkin-shaped seed capsule splits open when it's ripe, shooting its fragments and all 16 seeds up to a staggering

Hura crepitans

LATIN NAME

230 ft (70 m) across the rainforest. This "horticultural hand grenade" is quite a dramatic method of seed dispersal, but means that the plant doesn't need to rely on animals to distribute them far from home.

Don't monkey about with this tree

The tree's trunk looks rather terrifying, clothed with curious conical prickles, not unlike the studs adorning a leather jacket. Their purpose is to protect the tree from predators. In particular, it prevents monkeys from clambering up the trunks—as we know, they love seeds and nuts. The common name "monkey no climb tree" now makes a lot of sense, although I doubt a human could scale one, either.

Inhabitants of the rainforest who can fly don't hesitate to take advantage of the leafy habitat *Hura crepitans* offers, though. Bats tend to hang out in the *Hura* lair and, in the process, they pollinate the flowers. The male flowers resemble clusters of berries and have no petals. The female flowers look more demure and are produced singly. Both occur in a rather fashionable shade of maroon. Once pollinated, the fruits develop—changing through green and yellow and soon ripening into a lovely shade of rustic brown.

Hura crepitans is most commonly known as the sandbox tree, a nickname that originates from the custom of filling the seed pods with sand, for blotting ink. Don't tell the stationery obsessives though, they'll all want one. But this desktop use is probably the tamest of all—and the least likely to kill you. More on that shortly.

Uses good and bad

Hura crepitans' leaves are heart-shaped, paper-thin, and 2 ft (60 cm) long, allowing the tree to grow in even the deepest shade, and giving it an edge over its rainforest rivals. It also thrives in wet or dry soils. If only my tropical houseplants were this flexible.

The trees are occasionally used to cast welcome shade over streets in the tropics, but their durability is outweighed by the risk of the seed pods exploding. It's made a nuisance of itself in Tanzania, where it's classed as invasive, because it outcompetes native plants. The wood from *Hura crepitans* is used to build houses, although care must be taken when felling as the toxic sap can cause temporary blindness.

Some of its other uses are not for the faint of heart—it is involved in the preparation of tear gas, and the latex harvested from the bark is an important component of arrow poison for the Caribs of the Lesser Antilles islands. Gosh.

> # The trees are occasionally used to cast welcome shade over streets in the tropics, but their durability is outweighed by the risk of the seed pods exploding.

Hunters in the Amazon used to drop the toxic sap into streams causing the fish to surface for oxygen where they could be caught.

The spined trunk wards off potential predators looking to climb the tree and snack on the delicious seeds.

MISTAKEN IDENTITY

02

Some people only interact with plants to laugh at them—and all because they just so happen to look like something else. But plants have to do everything they can to attract pollinators, even if that means becoming experts in disguise and deception—and not always in the way you'd expect.

Darth Vader flower

THE FLOWERING MENACE

HOW BIG IS IT?
The plant can reach 10 ft (3 m) tall.

WHERE'S IT FROM?
Areas of Central America, in particular El Salvador, and as far south as Colombia. But it's much more common on Instagram.

WHAT'S ITS NATURAL HABITAT?
Humid meadows and soggy floodplains.

HOW'S IT POLLINATED?
By beetles and flies.

CAN I GROW IT AT HOME?
Yes, if you have the warmth and a tropical vibe.

Some plants get internet famous. Others don't. Back in 2017, a photo of a jumping dolphin plant (*Senecio peregrinus*) created a buzz when it got 10,500 retweets on Twitter because the plant's leaves look like dolphins. Others often posted online include Hot lips (see page 58) and the monkey orchid (*Dracula simia*), which has a flower that does indeed uncannily resemble a monkey's face.

But the most arresting of all cases of mistaken identity has to be the Darth Vader flower—and the pictures of it haven't been anywhere near Photoshop. This particular *Aristolochia* is quite a rare

Aristolochia salvadorensis syn. *arborea*

LATIN NAME

plant in the wild, but I was lucky enough to see one at the botanical gardens in Niigata, Japan, where I also met the telegraph plant (see page 118).

The full cinematic effect of the Darth Vader flower isn't apparent at first, as the creepy hooded blooms are held close to the base, almost hidden from human view. But, lo and behold, when I crouched down, I was soon staring right into the face of the once-heroic Jedi Knight. If I were a *Star Wars* fan, I'd insert a knowing quip here, so I apologize. I'm more likely to be found watching soaps than sci-fi.

Going all out to attract pollinators

Up close and personal, the first thing you notice is the odor of an overflowing trashcan, since rotting meat is the scent of choice for the plant to attract its pollinators. The back of the hooded bloom is even covered with maroon veins, in an attempt to look just like a slab of meat as any creature approaches.

Inside, things get murkier, as the hood casts a shadow. Staring out of the darkness are two luminous "eyes," whose job is to

Staring out of the darkness are two luminous "eyes," whose job is to entice pollinators into the deepest, darkest parts of the flower.

When inside the large flower, extra nectar bounties are often available to the insect—in case they get stuck a little longer than they'd hoped.

entice pollinators into the deepest, darkest part of the flower. Here, beyond the eyes, a wall of sticky hairs slows them down, dusting them with pollen, like an economy car wash. As the hairs soften, the insect is free to go—with the plant hoping it will toddle off to pollinate another bloom.

Stealing the show

In the wild, *Aristolochia salvadorensis* is a straggly, rather unkempt plant, weaving its way up and across its natural habitat. The foliage is pretty and wisteria-like, but with all the theatrics happening down below, it is usually overlooked.

When established, plants flower prolifically from their corky base. The fascinating flowers often stay open for a week, in which time they play host to an array of visiting pollinators. They are within easy stumbling distance for beetles to find, but they attract flies to their meaty platter, too. An invitation to a party in the dark? Yes, please!

COMMON NAMES

Darwin's slipper, happy alien plant

THE PLANT FROM ANOTHER WORLD

HOW BIG IS IT?
5 in (12.5 cm) tall, with a spread
of 10 in (25 cm).

WHERE'S IT FROM?
The craggy peaks of the Andes in Chile
and Argentina, as well as Patagonia.

WHAT'S ITS NATURAL HABITAT?
Rocky mountain slopes with thin soils,
but also close to sea level in the coldest,
southernmost parts of the plant's range.

HOW'S IT POLLINATED?
By herbivorous birds called seed snipes.

CAN I GROW IT AT HOME?
Technically, yes, but plants that are true
to type can be rare, unfortunately.

**What's that coming over the mountain?
Some say penguins ... others think they
look like aliens ...**

Whatever your belief, *Calceolaria uniflora*
is an evergreen alpine plant that keeps
itself tight to the ground, with shallow roots
and nest-like foliage. The plant's blooms,
though, are a talking point—orange-yellow
with red marks, chestnut freckles, and a
white band across the "mouth" (more on
that in a second).

The so-called alien plant certainly grows
in the right surroundings. The rocky, often
mountainous outcrops it resides on have
a really otherworldly look about them.

Calceolaria uniflora

LATIN NAME

It's far too cold here for bees and insects to survive and pollinate the flowers though, so what's a plant to do?

Well, researchers have observed birds grazing on the seeds and fruits of many alpine plants, including the blooms of *Calceolaria uniflora*. You see, the flower isn't designed to look like an alien to attract other aliens and have its flowers pollinated by them! This appearance is incidental.

Lip-smackingly good

The lower white lip of the flower is apparently rather tasty—almost almondlike, I am told. It's high in sugar and an essential part of the diet of the seed snipe, a gregarious wading bird. Groups of them zigzag around, devouring these white strips of goodness in record time, the greedy things. The red-and-yellow backdrop of the rest of the flower highlights this glorious buffet even more.

> The lower white lip of the flower is apparently rather tasty— almost almondlike, I am told.

Well, here's where things get pretty awesome. You see, it just so happens that the distance between the tip of the seed snipe's beak and the top of its head is $^3/_4$in (1.8 cm), exactly the same distance as between the white strip and the reproductive parts of the *Calceolaria uniflora* flower. As the bird tucks in down below, its head becomes doused in pollen. As it moves to the next "feeding table," its head brushes against the bloom and the flower is pollinated. A peculiar process indeed.

HOW CURIOUS

Charles Darwin found this plant on a voyage around South America in the 1860s—hence the name Darwin's slipper. *Calceolus* means "little shoe" in Latin.

The *Calceolaria* family tree

The alien plant is quite rare, so you may find it hard to track down any young plants, especially as the species is often sold mislabeled or on the illegal market.

Calceolaria is a rather intriguing family, and the alien plant has a relative with an amusing appearance—the balloon flower (*Calceolaria integrifolia* and other hybrids), which has inflated flowers in a rainbow of colors. It's a plant that's fallen out of favor with the patio crew, along with *Celosia*, *Cineraria*, and a few other untrendy plants that start with the letter C. But, developments are afoot. After all, we're all wearing '80s clothes, so why not grow '80s plants, too?

The goggling eyes of
Calceolaria uniflora are in
fact the reproductive parts
of the flower and carry pollen.

COMMON NAME

Flying duck orchid

HOW BIG IS IT?
8 in (20 cm) tall, with a spread
of 6 in (15 cm).

WHERE'S IT FROM?
East and South Australia.

WHAT'S ITS NATURAL HABITAT?
Sandy heathland, often beneath
eucalyptus trees.

HOW'S IT POLLINATED?
By (unsuspecting) insects.

CAN I GROW IT AT HOME?
Almost impossible.

We have a bit of a catfish here—this plant is pretending to be something else entirely, although not what you might think at first. And it isn't just its appearance that's misleading—this mischievous orchid takes deception to the next level in what I can only call some serious catfishing.

Despite what your eyes may be telling you, this plant isn't intended to look like a duck in flight. Nature has, in fact, designed it to resemble a female sawfly. And in more ways than one—the plant even secretes a chemical that's similar to the insect's sex pheromone. So, why does it bother going to all that trouble?

THE PLANT THAT'S A TOTAL QUACK

Caleana major

LATIN NAME

Duped into it

Well, with the sight and scent of a "female" right before him, the unsuspecting male sawfly falls into a first-date frenzy. With the flower's petals swept backward, the labellum (central petal) appears super enticing, so the male sawfly attempts to get jiggy with the flirty flower and engages in pseudocopulation (which needs no explanation).

Before long, his weight forces the labellum down, which traps the insect. The only means of escape for him is to edge out backward and past the pollen. I imagine it's harder for a sawfly to avoid the gold stuff in a *Caleana* flower than it is to dodge duty free at the airport.

After coating the insect, the labellum flips back into place. The eager Lothario then flies off to his next date, with his belly and wings covered in pollen, which he distributes onto another flower, unknowingly pollinating it with the utmost efficiency.

With the sight and scent of a "female" right before him, the unsuspecting male sawfly falls into a first-date frenzy.

Another twist in the tale

Then it turns out that the orchid is already "in a relationship," although this time it's the roots that are involved. They have a symbiotic interaction with a specific type of fungus that only occurs in the areas where flying duck orchids grow. The fungus protects the plant from other fungi and provides it with nutrients. In return, the fungus derives sugary goodness from the plant's roots. This is the reason

In 1986 the flying duck featured on a postage stamp in Australia—one of four stamps celebrating the country's native orchids.

that these orchids can't easily be grown in cultivation—the conditions are simply too hard to replicate. Although I do know of some examples, shrouded in secrecy in an unnamed botanic garden in Australia. The exact method of cultivation is information as highly classified as the recipe for Coca-Cola.

Hard for humans to spot

Caleana major flowers from September through January—that's summer in Oz. At only 8 in (20 cm) high, with thin, wiry stems, the plants are neatly camouflaged among the heathy understory of the Australian eucalyptus forests, often in a red-brown landscape, too, making them hard to find in the wild. Unless, that is, you're lucky enough to live in Melbourne, and you've got the time to go out searching for them.

Wild populations of *Caleana major* are sadly now marked as vulnerable. The destruction of habitat and loss of natural environment can clear plants, and this also means the pollinators—those poor male sawflies that seem so easy to dupe—are less plentiful, too.

Male sawflies partake in "pseudocopulation," in the process they brush pollen onto themselves that they then take to other flowers.

Hot lips, girlfriend's kiss

HOW BIG IS IT?
3–13 ft (1–4 m) tall: this plant is a shape shifter that can climb or form a bush.

WHERE'S IT FROM?
Central America and northern South America, including Colombia and Ecuador.

WHAT'S ITS NATURAL HABITAT?
The understory of rainforests. It's a sensitive plant that needs very specific conditions to grow.

HOW'S IT POLLINATED?
By hummingbirds, butterflies, and moths.

CAN I GROW IT AT HOME?
Gosh, no.

As well as making the plant a popular gift on Valentine's Day, the puckering **"lips"** of *Palicourea elata* have earned it names that range from the innocent girlfriend's kiss, to the slightly more raunchy hot lips. I've even heard it called Mick Jagger.

I've chosen the moniker "the most kissable plant," but you have to be quick! No sooner have the red lips appeared then they're joined by a fancy white flower, pushing its way through like a drag queen on the runway.

Those lips, you see, aren't flowers. They are bracts, which are modified leaves designed in bright colors to

THE MOST KISSABLE PLANT

Palicourea elata syn. *Psychotria elata*

——————————————————————— LATIN NAME

attract the crowds. Poinsettias—a favorite plant at Christmastime—have this same flower setup; they also have the most wonderful red bracts.

Just like human lips, *Palicourea elata* plants' come in a myriad of shapes and sizes. With no scent, and growing in the alleyways of the forest underworld, the plant really has to work hard to attract pollinators, but those kissable lips seem to do the trick, beckoning hummingbirds to them, along with butterflies and moths.

Once flowers are successfully pollinated, the plant produces a jet-black berry. It's thought that birds gobble these up, pooping them out somewhere else in the forest. Interestingly, plants can vary greatly in size and shape, depending on where they need to grow. If they're jostling for light with close contacts, growth is more elongated and almost climbing in habit, and the plants can be remarkably self-supporting. In more open situations, though, you'll see a bushier side to our hot lips.

Victim of its popularity

The plant has a dazzling array of medicinal beliefs attached to it—in Nicaragua, for example, it's used to calm the side-effects of snakebites. The efficacy of most treatments is anecdotal, however, and little or no medical research is thought to have been carried out. Plants are also said to have a psychedelic effect and were commonly used in ceremonies in times gone by.

In addition to the medicinal folklore, the sheer numbers of

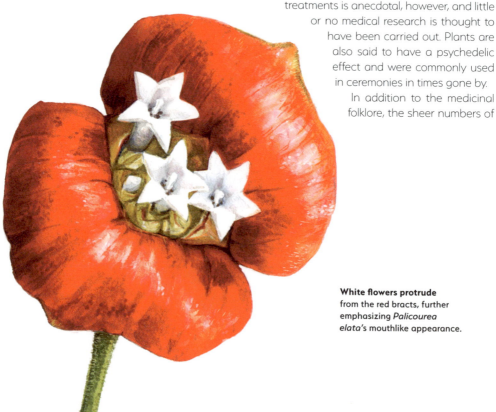

White flowers protrude from the red bracts, further emphasizing *Palicourea elata*'s mouthlike appearance.

Palicourea belongs to the *Rubiaceae* family, also known as the coffee family.

plants traded as tokens of love on Valentine's Day are taking their toll on *Palicourea elata*—these lips are crying out to be left alone. As well as this, the effects of climate change and loss of habitat are putting hot lips even more at risk. Since *Palicourea* grows in the understory of the rainforest, it needs forest-floor shade to thrive and survive and cannot exist elsewhere. Expansive farming, leading to fields being cleared for crops and pastures reserved for livestock, is putting pressure on the spots that hot lips likes to call home.

Those kissable lips seem to do the trick, beckoning hummingbirds to them, along with butterflies and moths.

Viral—in a nice way

So, if you're lucky enough to encounter hot lips in the wild, I suggest you leave it there. Take a pic instead and post it on Insta. Photos of these flowers always go viral, in the same way that one of *Senecio peregrinus*—a plant that grows tiny "leaping dolphins"—once did on Twitter in Japan, with upward of 10,000 retweets.

MISTAKEN IDENTITY

Dead man's fingers

HOW BIG IS IT?
An individual fruiting body is 1¼–3 in
(3–8 cm) tall, and 1½ in (4 cm) wide.

WHERE'S IT FROM?
Europe and North America, but
technically it can pop up anywhere.

WHAT'S ITS NATURAL HABITAT?
Deciduous forests and woodland,
usually in the base of rotted and
damaged tree stumps.

HOW DOES IT REPRODUCE?
Spores are spread by flies or
dispersed into the air.

CAN I GROW IT AT HOME?
Not easily—and it can infect trees with
little warning, causing black root rot.

Unlike the alien plants in *The Day of the
Triffids*, someone should write a movie
one day about real plants and fungi that
are just as terrifying. And there's none
more worthy of playing a starring role than
Xylaria polymorpha. Go off track in the
forest at the wrong time of year and you
might spot some blackened fingers clawing
their way up from the forest floor.
Don't panic, though, it's only a fungus
trying to live its best life.

Xylaria polymorpha can appear in many
handlike forms, from chubby to skinny, and
with or without all five fingers. It's positively
ghastly. This fungus has a cosmopolitan
distribution, which means that, given the

THE
CREEPIEST
OF ALL FUNGI

Xylaria polymorpha

LATIN NAME

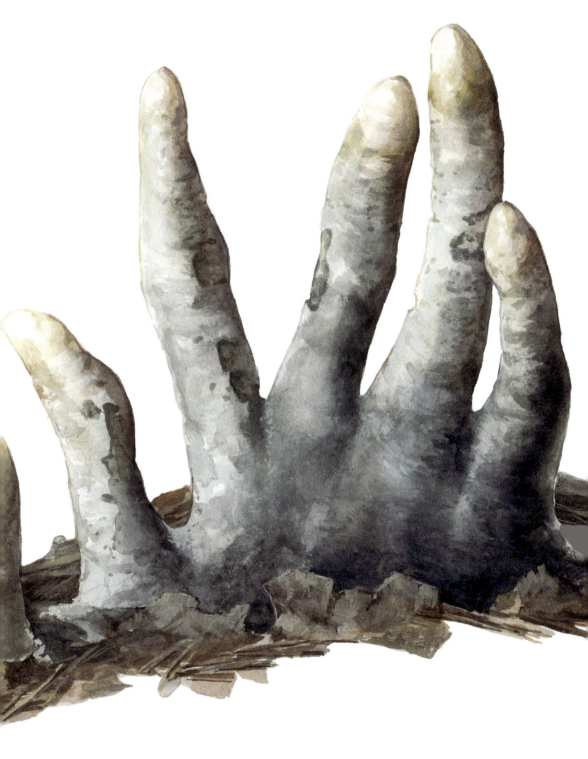

right habitat, dead man's fingers can appear just about anywhere in the world, including your backyard. In fact, the threadlike structures of the fungi, called *hyphae,* can spread vast distances underground and are quite difficult to detect. They could be right under your feet and you wouldn't even know it.

Plants versus fungi

Now, I wouldn't rap your knuckles if you referred to *Xylaria* as a plant. I'm not like that. But fungi are not really plants at all. They differ in many ways—and not only in their otherworldly appearance. Plants grow by using sunlight, carbon dioxide, and water in a process called photosynthesis. Plants also have chlorophyll, which gives them the green color we know so well.

Fungi appear rather different. They grow mostly out of sight and are formed of networks of hyphae that, as a whole, are referred to as a "mycelium."

Some fungi make spores—tiny, reproductive cells that can give rise to a new individual without any sexual happenings; unlike plants, which need a bit of jiggy. Mushrooms are the fruiting body of the fungus and their job is to distribute those spores.

> ## Dead man's fingers can appear just about anywhere in the world, including your own backyard.

A rotten diet

But back to the *Xylaria,* which feeds on decaying matter. We call that "saprobic" and it's almost like a public recycling service. The fungus breaks down dead organic matter into energy for the fungus, at the same time providing plants with the nutrients they need. To do this, dead man's fingers can sprout from the base of rotting stumps, in bark lesions, and even from the stalks of decaying herbaceous plants.

The fungus can invade damaged roots, too, leading to root rot and bringing about the death of the plant. So, sadly, despite being a curiosity, the presence of the fungus is an indicator of a tree with a problem, although apple trees have been known to produce one final crop before they die.

The fungus is also occasionally found living in and among termite mounds. It's thought that the termites take the spores back inside the mound stuck to their bodies. When the termites eventually die, the fungus gobbles up the partially digested materials they leave behind.

Changing appearance

In the course of its lifetime, *Xylaria polymorpha* takes on lots of different guises (*polymorpha* means "having many forms" in Latin). Young dead man's fingers are pale with white tips. In summer, they begin to blacken and, by late fall (coinciding neatly with Halloween), they're brown to black and embedded with spores. At this point, they are actually quite hard to spot among dark woodland—until they reach up behind you and scare the heck out of you, that is!

The fingers can last a year and keep releasing spores into the surrounding area, just waiting for the wind and the rain—as well as flies—to take the spores on their onward journey to a rustic growing spot.

HOW CURIOUS

The fungus prefers rotting elm, maple, beech, and apple trees. Once it has taken ahold, the trees can collapse and fall without warning.

COMMON NAME
Marble berry

THE FOREST'S HIDDEN GEM

HOW BIG IS IT?
18 in (45 cm) tall.

WHERE'S IT FROM?
Africa, from Congo to Angola.

WHAT'S ITS NATURAL HABITAT?
The forest floor.

HOW'S IT POLLINATED?
Hoverflies and bees pollinate some *Pollia* species and they are believed to help with this one.

CAN I GROW IT AT HOME?
Apparently, it's fairly easy to grow, but mega hard to get hold of.

It takes a bit of explaining to understand how this amazing plant shines so bright, so I hope you're sitting comfortably. First of all, it isn't blue, as it initially appears, and let me tell you why ...

The color—often referred to as the most intense of any biological material—is created by "structural coloration." On the surface of every tiny $\frac{1}{8}$in (4 mm) berry, a smooth and transparent cuticle reflects light, just as a mirror does. Inside the berries, many layers of cells are stacked up, but they don't bring any color. They simply reflect light from many different angles and it's these reflections that produce the different colors, with the spherical shape of the fruit encouraging and amplifying them big-time.

Pollia condensata
LATIN NAME

The end result is a glossy, bright fruit that shines purplish blue, changing from every angle and looking almost pixelated at times.

The end result is a glossy, bright fruit that shines purplish blue, changing from every angle, and looking almost pixelated at times. Very cool. And it's incredible to think that the reason behind the beautiful color was only discovered very recently. A specimen found in Ghana in 1974 was taken to Royal Botanic Gardens at Kew, in London, and its intriguing purple-blue color studied for many years.

The explanation for it finally came in 2012. It was a team at Cambridge University Botanic Garden, which includes several botanist friends of mine, who discovered

The surface of the berry reflects light in an infinite amount of directions, creating a glittering multicolored effect up close.

The "structural coloration" that creates the berries' iridescent effect, is also what creates the colors in a peacock's feathers.

that the multi-layered structural coloration of *Pollia condensata* is, in effect, formed by thousands of tiny mirrors. The intense color has 30 percent of the reflectivity of a silvered-glass mirror. It is the only plant known to have this feature.

From inconspicuous to inconceivable

The flowers that precede the marvelous berries are really quite inconspicuous. They are somewhat drab in style, color, and form, and blend into the undergrowth. Yet, once pollinated, they soon form the most complex of all berries. The hardy, dry, ever-sparkling fruits try their best to stand out on the dark forest floor. Amazingly, they can shine for many years, way after they've been picked from the plants.

The berries have long been used for decoration in West Africa, but in recent times, jewelery designer Maud Traon went a step further, combining the rare berries with metals and gems in an incredibly vivid palette of colors. The resulting cuff link was a one-off and priced at almost £10,000 ($13,000).

Attractive to birds

But it isn't just humans who have a penchant for the reflective beauty of the marble berry. Research suggests that the sparkly fruit evolved to "deceive" birds. Despite the berries having zero nutritional value, birds happily snack on them, wolfing them down, and then casually distributing the seeds when they next poop.

Beguiled by the fruit, the birds sometimes even adorn their nests with the bright baubles. How silly—I mean, who'd leave their Christmas decorations up all year?

Living stones

THE PLANT HIDING IN PLAIN SIGHT

HOW BIG IS IT?
¹/₂–1¹/₄in (1–3 cm) tall, with a width of 1–3 in (2.5–8 cm).

WHERE'S IT FROM?
South Africa, Botswana, and Namibia, from sea level to high mountains.

WHAT'S ITS NATURAL HABITAT?
Dry grassland; veld (open countryside); and bare, rocky ground (if you can spot them).

HOW'S IT POLLINATED?
By insects or a gentle breeze.

CAN I GROW IT AT HOME?
Yes, and I would recommend it. It's such fun to grow from seed.

Lithops are a fine example of plant mimicry, with natural camouflaging so good that different types are still being discovered to this very day.

These really are fascinating plants, which prefer to blend in, like many of us at the parties and networking events we're aching to sneak out early from. In the case of living stones, the reason they've evolved to look like pebbles and merge into their surroundings is to avoid becoming a tasty snack for a hungry herbivore.

Lithops consist of a pair of giant fleshy "blobs," which are actually the leaves. These unusual plants are primarily buried

Lithops species

below soil level, with only the top visible. The surface of each leaf features a "leaf window," which is partially translucent to allow light into the subterranean part of the leaf, so the plant can photosynthesize.

But these leaf windows aren't just produced in random colors. Each species of *Lithops* has evolved to imitate the rocky surfaces among which they grow. Even within grassland, they'll usually mimic the ground. The markings therefore vary greatly and that's what makes *Lithops* so collectible. As succulents, they are perfect for the beginner. Their thickened, fleshy growth retains water very efficiently in challenging environments Eight-year-old me had a fabulous collection of living stones, which I had mostly grown from seed. It was a joy to watch them progress from seedling to adult, much in line with my own puberty.

Eight-year-old me had a fabulous collection of living stones, which I had mostly grown from seed.

Do try this at home

Intrigued by living stones? Well, I already know you'll kill your first *Lithops*—but with kindness. The number-one mistake tends to be overwatering. You have to bear in mind that in the wild, most species of living stones rely only on dew for their moisture requirements. And they don't grow during the winter, so you can put your watering can back under the sink for a while. Leave the plant alone until the spring, when the whole thing will split open, with a new plant emerging from the center of the crevice. Keep sitting tight and, as challenging as this might sound, don't water them yet. The plants can survive in conditions way harsher than the posh luxury of our home windowsills, and minimal watering once a month in the summer will be ample.

A surprise within the disguise

Now, the biggest surprise to you will be the flower. Later in summer, the curvaceous leaves part, and a protruding bud takes charge. This will soon open into the most decorative, daisylike flower. It will generally be white or yellow, but orange and red ones are out there, too. Get up close and you may find that it's fragrant. The flower is followed by a dry seed capsule, which disperses in the rain—a technique most *Lithops* in the wild rely on for seed dispersal. When a raindrop lands on the capsule, the seeds shoot forth like a James Bond baddie from the passenger seat of a speeding Aston Martin.

 Lithops could be part of your next party trick. It's rather fun to match the plants to stones as closely as you can, then see if people can tell them apart. Once you do this, you'll also understand how new types of these curious little "pebbles" are still being found.

The name "Lithops" comes from the Ancient Greek, "líthos" meaning "stone" and "ops" meaning "face."

Come summer, *Lithops* flowers can be seen protruding through the pair of "skylight" leaves.

COMMON NAMES
Tulip orchid, cradle orchid

HOW BIG IS IT?
16–24 in (40–60 cm) tall, with a spread of 18 in (45 cm).

WHERE'S IT FROM?
South America—Venezuela and Colombia, mostly.

WHAT'S ITS NATURAL HABITAT?
The forest floor at high elevations, sometimes growing on mossy rocks or other plants.

HOW'S IT POLLINATED? Pretty much only by male euglossine bees.

CAN I GROW IT AT HOME?
Yeah, definitely, and I wish I had one.

Let me introduce you to one of my favorite flowers—the cradle orchid. It belongs to a plant family that is possibly one of the most adaptable in the natural world. Orchids have a greater knowledge of "the birds and the bees" than any postpubescent teenager ever could. They are overtly sexual plants, often with their reproductive organs on full display, doused in a seductive perfume to attract a range of animal and insect life, and they can take on a range of alter egos to get exactly what they want.

THE FLOWERS THAT ROCK MY WORLD

Anguloa clowesii
LATIN NAME

Anguloa clowesii, for instance, are not shy and retiring in the slightest. You'll find them on the forest floor at high elevations across countries from Venezuela to Peru. Their huge, theatrical leaves can reach up to 3 ft (1 m) in length. These are soon followed by two flower stems that rise from the fleshy, bulblike base, like the most beautiful tulips.

Color and scent

The flowers of the *Anguloa* genus are the most delightful pastel shades. *Anguloa clowesii* has lovely, buttery yellow blooms. Once open, their internal parts are fully exposed. These just so happen to look like a crying baby, flailing its arms around to get attention in its cradle. I love this plant for its high drama.

Remarkably, the 3-in (8-cm) flowers last many months, allowing plenty of time for visits from potential pollinators. As with many orchids, they have an intriguing fragrance that is important in attracting pollinators. The scent is a heady mix of coconut, peppermint, and chocolate aromas that would provoke discussion and arousal among any group of wine buffs.

This overwhelming scent is designed to lure its pollinators of choice—male euglossine bees. The outer lips of the flower act as a landing strip for the inquisitive bees, who can't resist crawling into the narrow opening to investigate further. When a bee happens to step on the lip of the flower, which floats above the reproductive parts, the bee's weight upsets the balance, causing the flower to hinge backward, like a cradle being rocked, and dump the bee into a bath of pollen. Once the bee is covered in pollen, it is then free to go on its merry way, hopefully to another *Anguloa clowesii* bloom, which it will then pollinate as it clambers inside, still dressed in its (very fashionable) pollen coat, and once again tips itself onto the flower's sexual parts.

The scent is a heady mix of coconut, peppermint, and chocolate aromas that would provoke discussion and arousal among any group of wine buffs.

The genus *Anguloa* was named in honor of Francisco de Angulo, a renowned collector of orchids in the 19th century.

A helping hand from me

Thanks to an epic display by the Orchid Society of Great Britain at Chelsea Flower Show a few years back, I got to see the cradle-rocking in action. There were no male euglossine bees to be found in the Floral Marquee that day, so it may well have been my hand that rocked the cradle, just to test nature's best booby trap.

Quite an imposing sight, with glossy leaves and indulgent blooms held on strong stems, *Anguloa clowesii* could easily be mistaken for tulips.

GREATER
GOOD

03

In our ultra-modern world, it's easy to forget where things come from, whether they're foodstuffs or raw materials. So, in this chapter we shine a light on some of the most commercially successful plants the planet has ever seen. Chances are your life wouldn't be the same without them, yet you know very little about them.

COMMON NAME
Tea plant

HOW BIG IS IT?
In the wild, 20 ft (6 m) tall, with a spread of 8 ft (2.5 m), but cultivated plants are usually kept at 3 ft (1 m) high, to make the tips easier to harvest by hand.

WHERE'S IT FROM?
Discovered between Myanmar and China, but native to other areas of Asia, too.

WHAT'S ITS NATURAL HABITAT?
Rich, moist locations. High elevations give tea the best flavor.

HOW'S IT POLLINATED?
By small flies, mostly.

CAN I GROW IT AT HOME?
Yes, anywhere you can grow a usual ornamental camellia.

If you're British and you're reading this, you'll know that a cup of tea has been scientifically demonstrated to be "the answer to everything." It's a belief that's shared by millions of people around the globe, making tea the most popular pick-me-up in the world, from a freshly poured *masala chai* in Mumbai, to a cold Hong Kong milk tea from the refrigerator in Mong Kok.

While these are both made from black tea, that's not the full story when it comes to *Camellia sinensis*. It is also harvested as green, white, yellow, and oolong tea, with different processing methods giving different results. Green tea, for example, is

GROW YOUR OWN TEA BREAK

Camellia sinensis
LATIN NAME

Picked leaves can be dried and processed in a variety of ways to make black, white, green, or oolong teas, for all your beverage needs.

made from young leaves, which are heated then dried. With less processing than some other teas, green tea holds on to more of the health benefits.

~~~~~~~~~~~~~~~~~~~~~~~~

At altitudes of 4,900 ft (1,500 m), the tea plants grow more slowly, but are prized for their more flavorful tips.

~~~~~~~~~~~~~~~~~~~~~~~~

Over the years, tea has been a kind of cultural and economic "glue," holding the world together and playing a significant role in linking East to West, thanks to the historic and present-day trade of it.

The earliest confirmed evidence of tea consumption comes from the 2nd century BCE, from leaves found in the mausoleum of Emperor Jing of Han, in western China.

Harvested by hand

In modern times, the harvesting of tea has proved difficult to mechanize, which is why the image of picking teams hard at work in the hills of tea plantations is so ubiquitous. It's said that the best tea comes from higher locations. At altitudes of 4,900 ft (1,500 m), the tea plants grow more slowly, but are prized for their more flavorful tips. To create a flat "picking table," their height is restricted to around 3 ft (1m). The most desirable parts are the uppermost pair of leaves and buds, which are snipped off once a week. They are packed with nutrients at this age and have a better flavor and texture than older leaves. Expert pickers can harvest 77 lb (35 kg) a day, which, after processing, results in 20 lb (9 kg) of black tea. No machine could ever match that.

Camellia sinensis is quite easy to grow at home if you have moist, slightly acid soil. It's a handsome plant with glossy foliage and fragrant, white, 1½in (4 cm) blooms in October, which should look familiar, as they are cousins of our popular garden camellias.

Making the perfect home-grown cuppa

So, how do you make the perfect brew from a tea plant you've grown yourself? Using your finger and thumb, pick the topmost two bright-green leaves and bud from each branch. Leave them to wither for 24 hours, then roll them with your hands to release the juices, transfer to a baking sheet, and place in a warm, humid environment for up to eight hours. Bake in the oven at 390°F (200°C) for 15 minutes to dry, lower the temperature to 265°F (130°C) for another 10 minutes until crispy and dry. Next, put the kettle on; pour the freshly boiled water over the leaves; and, once brewed, drink with milk or honey, as you prefer.

Talking of teatime, my favorite cake from Wales, bara brith, is made using a cup of tea. It is said to have been knocked into the mix by accident originally. Nowadays, the dried-fruit mixture is soaked in tea beforehand to give the cake its distinctive flavor. My last Welsh hill climb to find wild sundew plants was fueled by a huge hunk of bara brith, don't you know. It's lovely and oh so moist .... See, tea is great. It's almost enough to convert the coffee drinker in me.

There's a drink in Hong Kong called yuenyeung. Made from tea and coffee mixed together, it was invented in the 1950s to keep workers energized. It's now enjoyed by everyone.

COMMON NAMES ────────────────────────────

Cotton, Mexican cotton

HOW BIG IS IT?
6½ft (2 m) tall, with a spread
of 6½ft (2 m).

WHERE'S IT FROM?
Mexico, the Caribbean, and parts
of South and Central America.

WHAT'S ITS NATURAL HABITAT?
Semi-arid regions.

HOW'S IT POLLINATED?
By insects.

CAN I GROW IT AT HOME?
Yes, but you need a warm summer—
I'll keep my fingers crossed.

When I wear cotton, I feel invincible. Or, at the very least, a bit posh. As a fabric, it is valued around the globe for comfort, breathability (what a wonderful word), and general durability (although perhaps not when it's a white T-shirt and I'm indulging in a spot of impromptu gardening).

No surprise, then, that cotton is one of the most economically important plants on earth. Ninety percent of the world's harvest comes from *Gossypium hirsutum*, a handsome plant with an imposing stature. The foliage is palmate (hand shaped) and the flowers are silky and white-yellow, usually with a red eye. These soon fall to

THE MOST WEARABLE PLANT

Gossypium hirsutum
──────────────────── LATIN NAME

reveal chunky seed capsules. When cotton plants are happy and enjoying the warm weather, the ripening seed capsules burst forth with a flourish, revealing tufts of fluffy fiber, which almost obscure the inner seeds. At this stage, florists sometimes intercept the stems, such is their ethereal beauty.

The light fluffy material (lint) has evolved to help the seeds disperse in the wind and it's this part that's harvested. Like coffee and tea, cotton is still largely picked by hand. Done this way, the yield is greater than by machine and the cotton is also of a better grade. Longer fibers are preferred, and these are spun into yarns, then woven into all kinds of fabrics whose uses include making shirts, T-shirts, and undies (little did the cotton plant know that one day it would end up protecting the modesty of humans).

Not just for making fabric

But cotton isn't just grown for its fibers. The seeds certainly don't go to waste. Rich in oils and a valuable source of protein, they are widely used as an animal feed, although they are naturally toxic, so they need careful processing first. Roasted cotton seeds are sometimes ground and drunk as a coffee substitute, while cotton oil has various uses, from baking to skincare.

On their first day, cotton flowers are yellow-white, but once their pollen is shed, they turn reddish and begin to wither, before falling off a day later.

The typical cotton yield is 750 lb (340 kg) per 2½ acres (1 hectare) of land. That's enough to make about 300 T-shirt-and-jeans combos.

oil has various uses, from baking to skincare. Even the seedcake—the residue left behind after all the oil has been extracted—is turned into fertilizer.

Tainted reputation

But the true story of why cotton has been so successful is much more sinister, with a well-documented history of enslaved labor. While there are many organizations working hard to clean it up, the industry's notorious reputation for exploitative labor continues today.

As well as this, fertilizers and pesticides are still in heavy use in the cultivation of the crop—more chemicals are said to be sprayed onto cotton than any other crop. This has helped it survive competition from artificial fibers over the years, but there is a gradual movement to more environmentally friendly, and comfortable, materials.

> Little did the cotton plant know that one day it would end up protecting the modesty of humans.

While it's possible to grow your own T-shirt, you'll need to live in a warm climate. Cotton takes 100 days of good weather from planting to flowering, and a further 80 days until the browning capsules open to reveal the fibers. I guess I'll remain shirtless for now (the reader groans).

COMMON NAMES ——————————————————————

Cacao tree, cocoa tree

THE CHOCOLATE TREE

HOW BIG IS IT?
It can grow 15¾ft (4.8 m) tall.

WHERE'S IT FROM? Costa Rica and northern South America.

WHAT'S ITS NATURAL HABITAT?
The shady understory of the forest.

HOW'S IT POLLINATED?
By tiny flies—*Forcipomyia* midges, to be exact.

CAN I GROW IT AT HOME?
It's possible to grow a young seedling which could make a pretty cute houseplant. It's unlikely you'll be able to grow your own chocolate, though, unfortunately.

My favorite chocolate bar has always been a Double Decker. It is a melange of layered nougat and crispy cereal covered in the creamiest milk chocolate. Coming from the Greek words *theos* meaning "god" and *broma* meaning "food," the name theobroma literally translates as "food of the gods." And I am sure many of you would agree, as you curl up under your blanket on a Saturday night with a chocolate bar for company.

The cacao tree can seem slightly anonymous as it hangs out in the understory of the forest, but its flowers are dazzling, each one a cute little star. The

Theobroma cacao

—————————————————————————— LATIN NAME

These plants are "cauliflorous," which means the flowers bloom directly from the trunk and branches.

appear in, either as a result of the ripening process or the fact that they're a different variety. It's possible to see every shade from red through to yellow, with rustic orange tones in between.

Fruits ripen at different times on the tree, so harvesters cut them off with machetes, which they wield deftly to protect any developing flowers on the trunks. If they are too slow harvesting a ripe pod, the monkeys will get there first.

Trees are usually 4–5 years old when they start bearing fruits, and they thrive in a humid rainforest ecosystem. Sometimes they make themselves so at home that it's hard to distinguish between abandoned plantations and truly wild specimens. It's as if the plantations settle in so well that they become a part of the forest furniture.

All domesticated cocoa trees seem to originate from around 3,600 years ago in Central America and are now grouped into three main cultivars: Criollo, Forastero, and Trinitario. Criollo is the most prized and was the bean used by Mayan civilization. It is less bitter, with a more well-regarded aroma. Only 10 percent of chocolate is made from it, while 80 percent is made from Forastero, as it's generally a hardier and more disease-resistant crop.

blooms keep coming, all year round, and their diminutive structure suits pollination by small midges. Yet pollination remains stubbornly low because, just like people, some plants are just a little lazy and this plant is simply not that efficient. On one tree, 6,000 flowers across the year can result in just 20 pods.

Beans in a kaleidoscope of colors

Another rather fancy feature of the cacao tree is the kaleidoscope of colors the pods

Chocolate in times gone by

Theobroma cacao was first known to be grown as a domestic crop by the Olmec people in Mexico some time between 1500 BCE and 300 BCE, but the pulp was

The cacao was probably drunk as alcohol, with early mixologists combining it with maize, chile, vanilla, or honey.

In 1502, Christopher Columbus caught sight of a pile of cacao seeds from a canoe. Assuming they were almonds or goat poop, he paid scant attention to them, and failed to realize the potential they had.

probably a simple snack, not the cozy hot chocolate with marshmallows that you might be imagining.

Ceramic vessels with residues dating to 1900 BCE have also been found at archaeological sites, such as the Mokaya site in Chiapas, Mexico. The cacao was probably drunk as alcohol, with early mixologists combining it with maize, chile, vanilla, or honey, for ceremonial, medicinal, and culinary purposes.

Centuries later, Mayan culture made an unsweetened drink from the beans, but only for the elite. As Mayans migrated to northern South America and Mesoamerica, they set up some of the earliest plantations.

The first European to be introduced to the beverage was the Spanish conquistador Hernán Cortés in 1519. Within a century, the use of cacao spread across Europe, slowly reaching the masses.

Science to the rescue

Cacao seeds lose their viability when stored, so a "field gene bank" of living plants is the only option for preserving the genetics. As a result, some plantations effectively become "living libraries."

Climate change is driving the need for new varieties of cacao tree that can thrive in changing growing locations, and one such project is being run by the International Cocoa Quarantine Centre in Reading in the UK—sounds like a place you could go for some guilt-free gorging.

Today, large industrial plantations are still outnumbered by individual production by farmers with small plots. With careful harvesting required, growing cacao isn't easy to mechanize in the slightest. As such, it is still a very people-powered industry, hence the need for more visibility across the supply chain, and Fair Trade products.

In a wonderful eco move, the crushed shells are currently popular as an alternative mulch for gardens. The pulp from the chocolate-making process doesn't go to waste, either. It can be made into jellies, juice, or cream, and also fermented into an alcoholic beverage. To me, the thought of that drink has the same mythical allure that Irn-Bru has for some foreigners.

Coffee, Arabian coffee

HOW BIG IS IT?
In the wild, up to 39 ft (12 m) tall, with a spread of 8 ft (2.5 m), but cultivated plants are usually kept at 6½ ft (2 m) high to make the "cherries" easier to harvest by hand.

WHERE'S IT FROM?
Ethiopia, Sudan, and Kenya.

WHAT'S ITS NATURAL HABITAT?
The forest understory.

HOW'S IT POLLINATED?
Plants are self-pollinating.

CAN I GROW IT AT HOME?
You could give it a try, but the chance of producing cherries is slim.

I have six different ways of preparing coffee in my three-bedroom home. I can make espresso or percolate it in a machine. I can use a French press or brew Greek coffee in a stove-top *briki*. I can also make a glass of Vietnamese iced or a hot Japanese drip coffee. Whatever you want, but never instant. Not even the fancy cans. I am a snob like that.

Coffee is one of the most valuable commercial crop plants in the world. I hazard that much of that is fed by addiction. There are over 100 species, as well as some hybrids native to central

THE PLANT THAT'S FULL OF BEANS

Coffea arabica

LATIN NAME

Coffee geeks can be just as bad as wine buffs, finding the flavors and scents of different blends potentially orgasmic.

Before the cherries containing the coffee beans form, the plant bears clusters of white, jasmine-like blooms that are self-pollinating.

Africa, Madagascar, and tropical Asia and Australasia. Around 60 percent of global production is *Coffea arabica*, the species I'm writing about here. The remaining 40 percent or so is *Coffea canephora*, which has a less fruity, slightly bitter taste.

The first existing coffee species, *Coffea arabica*, is believed to have come from Ethiopia and was cultivated for consumption in both Africa and Yemen. As Yemeni trade spread in the early 18th century, production also began in Indonesia.

Coffea arabica was found to be a natural hybrid of *Coffea canephora* and *Coffea eugenioides*, which came about as a result of changing environmental conditions in East Africa a million years ago. The *arabica* species grows successfully above 2,000 ft (600 m), as well as in tropical environments.

Its flowers are highly fragrant and self-pollinating, although bees often help out. The blooms are produced in great profusion; so much so, in fact, that plants can often overload themselves, favoring flower production over their own health. So, in cultivation, the flowers and fruits need to be thinned by hand to save the plants from near-death.

As the flowers fall, the developing "cherries" are revealed—$\frac{1}{2}$in (1 cm) in diameter, each contains two seeds (beans). The cherries go through a kaleidoscope of colors—from green-yellow to red and then a deep, sexy maroon. Awkwardly, they ripen at different times, so hand-harvesting is preferred. There is also a yellow-fruited variety so knowing when to harvest must be a nightmare .... Anyway, to make

Wild Kopi Luwak coffee is said to be the tastiest in the world. The beans are consumed by wild civets whose poop is collected and processed. Anyone for a crappuccino?

harvesting easier, cultivated plants are usually kept to 6½ft (2m) high.

After harvesting, the cherries are squeezed to access the beans, which are then slowly dried on mats in the sun for a month or so to reduce their moisture content to about 10 percent, before they're roasted and ground.

Geek speak

In my experience, coffee geeks can be just as bad as wine buffs, finding the flavors and scents of different blends potentially orgasmic. Harvests from Java and Sumatra, in Indonesia, are the most prized—for their heavy body and low acidity. To give balance, they are often blended with higher-acidity coffees (usually originating from Central America and East Africa). Ask a coffee expert, though, and they'll probably describe all of that in a much more "floral" manner.

Adapting to a changing world

Despite having naturalized widely in Africa, Latin America, Southeast Asia, and China, coffee species are increasingly under threat. Deforestation affects coffee directly, as it's an understory plant. What's more,

coffee seeds do not store well, so the only way to preserve them is as living specimens. However, new cultivars are being bred to stand up to rises in temperature, longer drought periods, and excessive rainfall.

One kind of coffee that might be new to you (as it was to me) is shade coffee. Shade-planting is a more sustainable way of growing the good stuff, so you can expect to hear more about it in the future. It's a complex plantation of many plant species which means the coffee plants get more nutrients naturally—from debris falling from above, so they don't need fertilizers. Shade-grown plants can also live twice as long as their sun-grown cousins. They are understory plants, remember, so they prefer this dappled light. The shade also protects them from frosts. And, alongside this more sustainable approach, aficionados say that shade coffee tastes better, too.

Don't hold your breath if you're trying to grow your own coffee harvest at home, unless you can simulate the conditions they need to produce beans—those of a humid mountain forest. I'd stick to the little coffee houseplants you can buy, with handsome crepe leaves and a glossy windowsill presence.

COMMON NAME

Pineapple plant

HOW BIG IS IT?
About 3 ft (1m) tall, with a spread
of 1ft (30cm).

WHERE'S IT FROM?
The Paraná-Paraguay River
drainages and up into northern
South America.

WHAT'S ITS NATURAL HABITAT?
Tropical forest.

HOW'S IT POLLINATED?
In the wild, by hummingbirds; in
cultivation, by hand.

CAN I GROW IT AT HOME?
Yes, slice the top off one and get
it planted today.

**Pineapples have been involved in all
sorts of dramas over the years,** from being
the subject of aristocratic rivalries to
getting caught up in the drug trade.
As the third-most-important commercially
grown fruit in the world (at the time of
writing), it's been at the forefront of culinary
creation for years, and works really well
with ham on top of pizza (you can fight me
on that).

Although some archaeological evidence
exists, little is actually known about the
pineapple's voyage into cultivation, but by
the late 1400s it was a staple food of
indigenous American communities and

A STATUS
SYMBOL

Ananas comosus

LATIN NAME

widely distributed. The first European to encounter these ever-exuberant fruits was Christopher Columbus, in Guadeloupe in November 1493, on his second voyage to the area. He took them back to Spain, from where they spread into the rest of Europe.

By the 17th century, pineapples were seen as a symbol of luxury, but seldom eaten. Instead, they were used as decoration—on dinner tables, for instance, and depicted in stonework, bedposts, and pineapple-shaped teapots, as well as being emblazoned across wallpaper, tea cloths, and napkins. Little did the aristocracy know that all these years later we'd still be using them for motifs on notepads, rulers, and as an emoji (as a fruit, you really know you've arrived when you have an emoji).

In 1661, Dorney Court in Buckinghamshire, was the first place in Britain to grow pineapples, igniting a rivalry between stately homes to cultivate what was then a mythical fruit. Pineapples were a symbol of the owners' wealth, and also testament to their gardeners' skill and experience.

When I was younger, pineapples were *the* exotic fruit to have. Alas, only at Christmas, though, and always from a can.

Chelsea Physic Garden, in London, soon upped the game with the installation of pineapple stoves. These small, glass-walled frames surrounded holes in the ground, in which potted pineapples sat, surrounded by horse manure and bark. The right combination triggered a fermentation process that gave off enough heat to bring the pineapple to fruit.

A Christmas treat

When I was younger, pineapples were *the* exotic fruit to have. Alas, only at Christmas, though, and always from a can. My parents obviously found the fresh fruit a little too extravagant, or maybe they had no idea how to extract the golden flesh. The pineapple rings were ejected from the can, duly cut into squares, and threaded onto cocktail sticks between cubes of cheese, then added to the festive buffet, usually poked into an upturned cabbage half that had been wrapped in foil.

Now, grab a notepad and pen—I'm going to help you win your next game of trivia. Pineapple fruits are actually a "collective fruit," composed of many berries that grow and fuse together. Up to 200 flowers can make up each fruit and, as they turn to berries, they join together (become coalesced) in interlocking helices.

The ever-purposeful pineapple

While we think of the pineapple as a fruit to be eaten, there's actually a huge market for the fibers contained in the strap-like leaves. In warmer climes, pineapple fiber can be produced with the minimum of energy, so is absolutely seen as more sustainable than

In general, pineapple fruits display three sets of spirals on their peel, thanks to the roughly hexagonal pattern of their scales. Check that out next time you're at the supermarket.

synthetic materials. It can be transformed into linen, mats, and bags. In the Philippines, it was always the traditional fabric of choice among the elite.

The wholly innocent pineapple can also become embroiled in the illegal drug trade from time to time, as boxes and crates of the fruits are often used as cover for some other rather more illicit cargo.

Growing pineapples

Pineapple cultivation is intensive work. Fruits need almost a year and a half to reach maturity, and pollination has to be done by hand in those places where the natural pollinators aren't present. Wild pineapples are sometimes harvested, too, but if they're anything like that houseplant, the pygmy pineapple, then I'd rather not. They are bitter and better admired in a "shelfie" photo than eaten.

You can grow a real pineapple at home, though. Fruit won't be guaranteed, but it will keep the kids entertained for all of an hour. Cut out the top rosette from the fruit, then peel away the leaves to reveal the stem. Trim the base to just below the leaf scars, then pot up in quick-draining potting soil, water well, and wait. Maybe growing your own will help you kick that Hawaiian pizza habit ....

Who knew the pineapple flower resembled the fruit so closely. The fruit itself is made up of a bunch of berries joined together.

COMMON NAMES ————————

Peanut, monkey nut, groundnut

THE ORIGINAL BAR SNACK

HOW BIG IS IT?
It can grow 12–20 in (30–50 cm) high with a spread of 36–48 in (90–120 cm).

WHERE'S IT FROM?
Areas of South America, originating, it's thought, in the region around the Argentinian and Bolivian border.

WHAT'S ITS NATURAL HABITAT?
The tropics and subtropics.

HOW'S IT POLLINATED?
Plants are self-pollinating.

CAN I GROW IT AT HOME?
Yes, and it's fun.

Where do we start? In all their many different guises, peanuts are a well-loved snack around the world. When I was a kid, and my dad used to sneak me into the bar at his social club, only a packet of salt-and-vinegar peanuts would keep me happy. Then, at Christmas, my eight-year-old self felt incredibly fancy at being given a pack of monkey nuts to shell. Shamefully, however, I didn't discover peanut butter and jelly sandwiches until my twenties.

So, given my penchant for peanuts, you can imagine my excitement when, at the age of 35, I was traveling by taxi in Japan. As we passed fields of golden rice, I spotted a swathe of lush green foliage, and small piles of brown "droppings." I knew what it

Arachis hypogaea
———————————————————— LATIN NAME

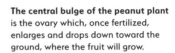

The central bulge of the peanut plant is the ovary which, once fertilized, enlarges and drops down toward the ground, where the fruit will grow.

was immediately and politely asked my driver to screech to a halt so I could investigate a crop of peanuts, freshly harvested and drying on the ground.

Not a nut at all

It's always rather a surprise when you find out how peanuts grow. The plants are geocarpic, which means the fruits develop underground. The yellow, sweet-pea-like flowers last only a day and, once self-fertilized, the base of the ovary elongates to form a "peg" that pulls it down into the ground. The fruit then forms in its own subterranean kitchen garden. Shock horror—peanuts aren't actually a nut at all. Like lentils and peas, they are in the legume family, hence the podlike shell they grow in. Remember that for the next time you do bar trivia.

As peanut plants are harvested, they are turned over and left to dry—just as I'd seen from the taxi on that quiet Japanese lane. They are then "threshed" (bashed about a bit) to remove the pods from the plants.

Different types for different uses

The modern-day monkey nut is a world apart from its wild cousins. Over the years, many different qualities have been selected: shape, size, flavor, oil content, and disease resistance. The different types of

> # I'm not sure where I'd be without peanuts. I'm a sucker for satay-slathered chicken.

nodules, which are home to symbiotic bacteria that produce nitrogen compounds to help the plants grow and compete with others. When the plants die, the nitrogen from them is then available to other plants and it also helps fertilize the soil. Win win, I'd say. Lupins and clover do the same thing. Write that down on your notepad.

Loved the world over, especially by me

The ongoing popularity of peanuts is such that world peanut production is an enormous 54 tons (49 metric tons) per year. At the time of writing, China is the biggest producer, growing 36 percent of the world's annual yield.

I'm not sure where I'd be without peanuts. I'm a sucker for satay-slathered chicken, I eat peanut butter straight from the jar (but with a spoon—I am classy, after all), and one of my favorite takeout dishes when I lived in China was a foil dish of mixed boiled soybeans and peanuts. Thank you, peanuts, you are my friend.

peanut grown are classified into groups. For example, the Spanish group is favored for its oil content, while the Valencia group is preferred for boiled peanuts. Most of those found in packets or on the bar counter are likely from the Virginia group, as they are the prettiest, whereas the Spanish make the best peanut butter.

Peanut plants are also valuable in crop rotations, because of their nitrogen-fixing properties. The plants' roots are covered in

The first commercially produced peanut butter in the world went on sale in the United States in 1895, but long before that the Inca of South America were grinding peanuts to make a tasty butter.

COMMON NAME ———————————————

Rubber tree

THE TREE WITH SOME BOUNCE

HOW BIG IS IT?
They can grow to 141 ft (43 m) tall, but cultivated trees are usually much smaller, as drawing the latex reduces growth.

WHERE'S IT FROM?
The Amazon basin and the Guianas.

WHAT'S ITS NATURAL HABITAT?
Tropical or subtropical rainforests with a yearly rainfall of at least 47 in (1,200 mm).

HOW'S IT POLLINATED?
By small insects such as midges.

CAN I GROW IT AT HOME?
How much space do you have?

Whatever you're into, rubber is a large part of our day-to-day lives and has been processed by human beings since as early as 1600 BCE, when early Mesoamerican cultures produced containers and recreational balls from it, as well as incorporating it into medicines and rituals. Reports from Mexico describe people dipping their clothes into latex to make them waterproof and more durable—think of this as an early form of windbreaker.

But before I get all technical, let me state how handsome rubber trees are and quite different from the ones we grow as houseplants (let's face it, you're unlikely to

Hevea brasiliensis

LATIN NAME

In 1895, the continuous tapping method for harvesting the latex from rubber trees was discovered.

have space for a true latex-producing tree). You have to look up quite a way to enjoy the canopy of fresh green foliage, among which pungent flowers nestle. Once the blooms have been pollinated, cute leopard-print seeds emerge from the ripening pods, exploding out much like the rubber tree's close relative, *Hura crepitans* (see page 40), but a little less noisily because these seeds are smaller.

Initially, rubber trees were only grown in the Amazon rainforest and, in an attempt to preserve the trade price, everything was done to protect the crop—even the export of seeds and seedlings was banned. But in 1876, 70,000 rubber seeds were hidden in banana leaves and smuggled to England, from where they were sent to Malaysia. Within 12 years, the plantations there had become as competitive as those in Brazil.

The continuous tapping method

In 1895, the continuous tapping method for harvesting the latex from rubber trees was discovered by Henry Nicholas Ridley, a scientist at Singapore Botanic Gardens. It's the process still in use today. Tappers peel back narrow strips of bark in a downward half spiral on one side of the tree. This "wounding" disrupts the line of the plant ducts, which release their milky latex. This latex, which is usually present so the tree can seal itself after injury and protect itself from insects, flows down the grooves and into a nifty collecting cup hung from the tree. After harvesting latex from one side, the opposite side of the tree is also "tapped," allowing the first side to heal. The work is carried out during the cooler part of the day, so the latex drips longer before it coagulates to seal the cuts. Each yield lasts around five hours. Tapped trees generally grow shorter, as the tapping unsurprisingly disrupts their growth to a degree.

The curious seeds of *Hevea brasiliensis* are toxic until processed. Processed seeds have been known to be used as famine food.

Hevea **flowers don't have petals,** but rather five triangular lobes. These are shorter than petals and sweetly scented to attract pollinators.

When the latex has been harvested, it is left to coagulate completely before being filtered and packed into drums. Acid is added to make the mixture clumpy, and it's then rolled into sheets and the water is squeezed out. Finally, the sheets are treated with chemicals and a low heat to become the black rubber we are all familiar with.

An industry bouncing back

In the 20th century, blights put pressure on Brazilian production. With very little genetic variation, *Hevea brasiliensis* had never built up a defense against plant diseases. Chemical protection and biological control have both failed to stop blights, and international freight and air travel increases the risk of diseases spreading. The good news is that, for now, the blight is isolated to that region and hasn't reached Asia yet—where most of the world's rubber production now takes place—as indeed you won't find many direct flights from South America to Asia. Plantations there have become more sophisticated in recent years with the trees usually being replaced after 20–30 years and the wood being used to make furniture afterwards.

However, new sources of rubber are increasingly important. The Russian dandelion (*Taraxacum kok-saghyz*) has been earmarked as a possible alternative, since the plant's roots are an excellent source of rubber. Watch this space.

HOW CURIOUS

About 20,000 types of plant produce latex which, in the event of injury, prevents bacteria and viruses getting inside the plant. The latex from only a few of them can be used to produce rubber.

COMMON NAME

Cashew tree

HOW BIG IS IT?
46 ft (14 m) tall and wide.

WHERE'S IT FROM?
Northern South America, including
Colombia, Ecuador, Peru, Venezuela,
and Brazil, and some Caribbean islands.

WHAT'S ITS NATURAL HABITAT?
Anywhere soil is fertile and
humidity high.

HOW'S IT POLLINATED?
By bees, ants, butterflies, and, less
often, by the wind.

CAN I GROW IT AT HOME?
I think it's a worth a try, but I can't
promise nuts. Be careful of the
shells' toxins, too.

How do you think cashew nuts grow?
If you're a bar trivia pro, skip this bit—we
don't want you spoiling the surprise.

Imagine a tree, ever so exotic, growing
like a candelabra with handsome leaves.
Clusters of flowers, each with around
500 blushing-red florets, adorn the plant.
These clusters open in a choreographed
formation: the blooms at the tips of the
flower sprays are hermaphrodite and open
first, with the males along the side stems
following later. The hermaphrodite flowers
(with both male and female parts) are a
fabulous food source with tons of tasty
nectar to keep would-be pollinators coming
back for more. The male flowers contain

A TOUGH NUT
TO CRACK

Anacardium occidentale

LATIN NAME

There's great peril for humans in extracting the nuts from the shells.

more sugar than the hermaphrodite flowers, which leads the pollinator into the addiction zone. In a feeding frenzy, the pollinator then visits all of the blooms, increasing the probability of pollination, just the way the plant likes it. All the blooms give off a sumptuous perfume, and

pollinating bees in particular can't get enough of the exotic nectar bounty.

Clusters of slender fruits are produced next, with awkward-looking growths the shape of a kidney bean attached to their tips. The fruits (known as cashew apples) get bigger and bigger, and their color changes from green to red as they ripen.

Soon, the curious lower growth is overshadowed by the cashew apple. But that grumpy-looking "kidney bean" is actually the star of the show. Inside is a hard double shell that encases the most flavorful of all nuts—the cashew. Technically known as a "drupe," this is the true fruit of the tree, and houses the seed. So move over, cashew apple—we are about to crack the case of the cashew nut.

The Tupi people, of Brazil, first realized there were nuts inside the drupes by watching capuchin monkeys access them successfully. But the tree doesn't give up its nuts easily. There's great peril for humans in extracting the nuts from their double shells. The outer shells need to be roasted off the

From flowering until nut maturity takes around 70 days and, even with pollinators, the success rate is only around 30 percent—some plants are simply not very efficient.

The world's largest cashew tree is in Brazil. Seventy times the usual size, it covers an area equivalent to two soccer pitches.

nuts in a well-ventilated area, as the edible part is surrounded by a liquid that's toxic to humans, although monkeys appear to be unaffected by it. The inner shell is then cracked and removed, revealing the tastiest nut of all (in my opinion, anyway).

But what happens to the cashew apples, the mere accessory in this situation? Alas, the fruit ferments quickly after picking and is too tender to transport, so it's seldom eaten outside the regions where it's grown. Ironically, it was the cashew apples rather than the hidden nut that first caught the attention of Portuguese explorers in South America. The kidney-bean-shaped parts were overlooked, as they were feared to be poisonous. Little did the explorers realize that there was a way of getting at them safely. They should have paid more attention to the capuchin monkeys!

My favorite nut

In my eyes, cashews are probably the sexiest of all nuts. When you bite into one, it just creams in your mouth. As a child, I remember being fascinated by the flavor and liked to coat my whole mouth with the foodie goodness. It was a revelation, especially compared with "bitter" walnuts, hazelnuts, and the like. However, they were the most expensive of all nuts, so they stayed a "Christmas only" treat.

Growing cashew trees as a crop

Cultivated widely in the tropics, the cashew is an accommodating tree, coping well on sandy soils and resisting drought. Dwarf forms are the most profitable, as they start producing in their first year.

A lack of plantation diversity, however, is now causing a decrease in the number of natural pollinators, and this is affecting the yield. Cashew farmers are being encouraged to plant colorful pollinator-attracting flowers alongside the trees and to avoid plowing around the crop, as this destroys bee habitats.

A multitude of uses

In the 1560s the Portuguese took the cashew to Goa, from where it spread further into Southeast Asia and Africa. Not only did Indian people discover some healing properties, they also created some truly delicious curries with the nuts.

The nuts are also made into butter (like a fancy peanut butter) and a hummus-like cheese and can be processed into a salad dressing and cooking oil. The 21st century has seen an increased demand for cashew nut milk. Does it make a good cappuccino? The jury is still out on that ...

SUPER-
HEROES

04

These are the record breakers, the most momentous of all plants, those we celebrate for being the largest, the smallest, or oldest. There's even one that I can only consider the luckiest. Plus, you'll meet a few that can do quite remarkable things, and I'll also tell you about a superhero plant that I created myself.

COMMON NAME

Egg and Chips® plant

HOW BIG IS IT?
4–5 ft (1.2–1.5 m) tall, with a spread
of about 12 in (30 cm).

WHERE'S IT FROM?
A development house in the
Netherlands.

WHAT'S ITS NATURAL HABITAT?
A greenhouse.

HOW DOES IT REPRODUCE?
This doesn't fit into an easy
category of reproduction, as it is
created by grafting two plants together.

CAN I GROW IT AT HOME?
Yes, and I would positively
encourage you to try.

For the bulk of my career, I worked in a role not unlike the Nutty Professor's, at seed and plant company, Thompson & Morgan. My job was to launch—and sometimes even create—brand-new plants. I worked on some pretty cool things, from tree lilies (unique types that grow taller than the average person) and climbing petunias (a concept never seen before), to blue verbascum (they're usually shades of pastel) and black hyacinths (any flower this color is chic by definition). They were sometimes hare-brained schemes, but they always made the headlines. Growing plants and gardening should be fun, after all.

A CROP WORTH GRAFTING FOR

Solanum tuberosum + Solanum melongena

LATIN NAME

At some point, I became a little bit obsessed with the *Solanaceae* family—potatoes, tomatoes, peppers, and eggplants. Around that time, "grafted" tomatoes were being trialed to give better crops on stronger rootstocks. Grafting involves joining two plants together, but they have to be compatible (from the same family) and the plants must be held together when young by small clips or even sticky tape.

Ketchup 'n' Fries

At Thompson & Morgan HQ, we held a few meetings to discover whether we could bring the concept of a TomTato® (a tomato and a potato joined together) back to the masses. We began a partnership with Beekenkamp BV, a Dutch company who had oodles of experience in grafting together tomatoes, or cucumbers, or melons. However they had never combined two entirely different plants—and that's what we'd need to do to make a TomTato®.

When we put the idea to them, they were doubtful, but they soon started experimenting and, after a few seasons of tests and trials, came up with a "duo plant," which gave super-sweet cherry tomatoes and a late-cropping potato. The potato had to be late cropping, otherwise you'd be digging up the potatoes before the tomatoes were harvested.

We launched the plant and the press went wild. I did tons of news interviews and was dubbed into many different languages. In the US, the plant was sold with the very descriptive name of Ketchup 'n' Fries, which you might still be able to find for sale on the internet, if you're interested.

Hard graft to make a new plant

My notepad and pencil went into overdrive. What else could be grafted? Perhaps we could make a roasted-vegetable plant with tomatoes and peppers? Or maybe a flower from that same family growing on top of a potato plant?

In the end, we decided to combine a potato with an eggplant. To my knowledge, "Egg and Chips®"—as we decided to call the plant—had never been served to a British gardening public before.

Eggplants are slightly more demanding to grow than tomatoes, though, so tests included trialing some of the latest eggplant varieties, which were known to crop well in lower light levels. Once we found a potato match, we were good to grow (excuse the pun).

Most importantly to me, this "Frankenstein" plant really caught the attention of teenagers—I could tell by the retweets.

Again, we hit the headlines. Among a flurry of tweets, radio interviews, and news clips, I was asked to appear on *The One Show*, the BBC live magazine program, to do a big reveal of the Egg and Chips® plant to presenter Chris Evans and actress Sarah Parish—all while wearing a burlap toga (it's a long story—when plants get famous, things can get weird).

Most importantly to me, this "Frankenstein" plant really caught the attention of teenagers—I could tell by the retweets. A new frilly fuchsia simply would not have given them the same excitement, and it proved to me that, to whip up interest in horticulture, you need to employ very different methods across different demographics.

When you see it, the Egg and Chips® plant is a bit of an illusion, as you might easily mistake it for a simple eggplant plant. But, the real magic is below the ground, as a bumper potato harvest is just waiting to be lifted. The plants are easily grown in containers and will crop reliably, even in a less than predictable British summer.

So, what dishes could you make with this vegetable all-in-one? Moussaka—a classic Greek dish of potatoes, eggplant, and lamb—would fit the bill perfectly. Or maybe you could whip up some of the smoky Middle Eastern dip baba ganoush, made from roasted eggplant, and serve it with chips—all grown by yourself.

HOW CURIOUS

During World War II, tomatoes were grafted onto potatoes as a nifty way to maximize space in vegetable gardens.

Telegraph plant, semaphore plant

HOW BIG IS IT?
It can grow up to 5 ft (1.5 m)

WHERE'S IT FROM?
Cambodia, China, and Sri Lanka, to name but a few places. Plants have even been found on the remote Society Islands, in the South Pacific Ocean. Maybe they did the conga over land and sea to get there ...

WHAT'S ITS NATURAL HABITAT?
It's an understory shrub, and it likes damp, bright conditions.

HOW'S IT POLLINATED?
We think by insects, but we're not sure.

CAN I GROW IT AT HOME?
Yes, it's easy from seed.

Ever imagined a plant you could sing to and it would dance? Well, the dancing plant is capable of rapid movement, easily visible to the naked eye. At last, something (rather than someone) that appreciates my passion for B-sides by failed 1980s artists.

First described as far back as 1880 in *The Power of Movement in Plants* by Charles Darwin, *Codariocalyx motorius* is the most legendary plant since *Jack and the Beanstalk*. With its flexible foliage, this captivating shrub can track the sunlight, following it throughout the day.

To do this, it has a pair of leaves that act just like nifty little solar panels. These smaller leaves constantly move elliptically

THE GREATEST DANCER

Codariocalyx motorius syn. *Desmodium gyrans*

LATIN NAME

to find the best angle for the larger leaves to face, feeding signals back to the plant—hence the name the telegraph or semaphore plant. The swelling and shrinking of the motor cells at the hinge of the leaf cause the "dancing" movements.

But the dancing plant's leaves don't waste their time moving to soak up the sun for no reason—that would be too much like hard work. As you might remember from school, light plays an important role in photosynthesis—captured light energy is used by the plant to convert water, carbon dioxide and minerals into oxygen and energy. What better way to maximize your light source than by moving yourself into it, something most plants don't have the luxury of doing.

~~~~~~~~~~~~~~~

These smaller leaves constantly move elliptically to find the best angle for the larger leaves to face, feeding signals back to the plant.

~~~~~~~~~~~~~~~

Voice recognition

Interestingly, telegraph plants have also been known to move for other reasons, one of which is in reaction to vibration. I've witnessed this for myself as I stood casually chatting next to one at Niigata Prefectural Botanical Garden, in Japan, and saw the resonance of my lively conversation seemingly cause the foliage to rotate. I am currently growing a few dancing plants from seed at home and it's my intention to try singing (badly) to them, in the hope they'll dance. I'll be the *cantor* and they will be my flamenco dancers.

Not in harmony

You may be startled to learn that there is no unanimous decision on why the plant does what it does. There's a school of thought that believes the plant may move to deter predators and that the visible movement of the leaflets could be aping the fluttering wings of a butterfly, for example. Except for the fact that the leaves droop when there's no light, which does suggest that light is the genuine reason for the movement.

If you're looking to accelerate the speed of growth of any plant, playing music, singing, or simply mouthing "you're growing well," is always worth a try. Various experiments have been carried out over the years to support this.

As I write this, my 3-in- (8-cm-) tall *Codariocalyx motorius* shows little sign of jiving yet, but I fear the climate in this box room is quite different from that of its natural habitat in South Asia. Nevertheless, I shall keep singing along to my favorite tunes on the radio.

HOW CURIOUS

Vibrations can actually stimulate the growth of any plant. Their preferred frequency is 115–250 Hz, which tends to be classical or jazz music, as well as the female voice.

I am currently growing a few dancing plants from seed at home and it's my intention to try singing (badly) to them, in the hope they'll dance.

COMMON NAMES

Resurrection plant, false rose of Jericho

HOW BIG IS IT?
2 in (5 cm) tall with a width of 8 in (20 cm).

WHERE'S IT FROM?
The Chihuahuan Desert in Mexico, but also in areas from Texas south to Costa Rica and El Salvador.

WHAT'S ITS NATURAL HABITAT?
Kicking around the desert floor, hoping for another burst of rainfall.

HOW DOES IT REPRODUCE?
It's a sporophyte, so plants are propagated by spores, rather than seeds or flowers.

CAN I GROW IT AT HOME?
Yes, and it's really fun for kids.

What a fantastic plant *Selaginella lepidophylla* is. Once it hits a water source, it can literally come back to life in front of your very eyes. One moment it's a dried-up piece of fern, the next a glorious green rosette of healthy growth. And, in the interim, it just blows around like a tumbleweed, as it auditions for the next big Western film.

The resurrection plant has a lot to cope with in the arid lands of central Mexico, one of the world's harshest habitats.

THE PLANT THAT CAN COME BACK FROM THE DEAD

Selaginella lepidophylla

LATIN NAME

Curled up like a tumbleweed,
Selaginella lepidophylla can be
blown around the desert undamaged
until it finds a new water source.

Most species couldn't make it, but this clever thing can lose 95 percent of its water without any damage at all to the plant. Once it finds water, even if it takes many years, it will usually fully recover—not just rehydrating, but switching photosynthesis back on, too. The plant literally comes back from the dead.

So, how does it find a water source? Well, when drought conditions persist, the roots can detach, so the plant can get blown around, searching for that drink. None of the frond-curling and folding as it tries to rehydrate seems to do it any damage.

But it isn't always so easy: if the plant is taken by surprise and doesn't have time to prepare, then resurrection can fail. Likewise, if the moisture-drought cycle occurs one too many times, the plant can't take it any longer, and dies. On the whole, though, these plants must be doing something right—they have existed for millions of years.

Use by missionaries

Selaginella lepidophylla is a dryland species, native mostly to the desert of Chihuahua, in Mexico. As the plant can seemingly come back from the dead, it was used by Spanish missionaries there to illustrate the concept of rebirth to their potential converts. It was often seen as a lucky charm and passed down from generation to generation.

Over the centuries, there have been various medicinal uses for this fascinating plant. For instance, there's an old wives tale that to induce childbirth you should drink the water the plant has soaked in—the speed at which the plant carries out its own rebirth is said to indicate the speed and ease of the actual child's birth.

Your new party trick

This is a fun plant to grow at home, where you'll be able to witness the resurrection right in front of your very eyes as plants go from "dead" to "alive" in around an hour. You should be able to find reputable sources for "dormant" plants online. Once the fern has resurrected, and looks like a conventional leafy house plant, you can pot it up into soil.

The resurrection plant is sometimes called the false rose of Jericho and is not to be confused with the true rose of Jericho, *Anastatica hierochuntica*. This tumbleweed plant can also expand and retract, but it cannot detach its root system or regain its green coloring, as the king of funky plants, *Selaginella*, can.

HOW CURIOUS

Resurrection plants have been known to survive extreme dehydration, and even desiccation, for up to seven years.

> This clever thing can lose 95 percent of its water without any damage at all to the plant.

COMMON NAMES
Miracle berry, miracle fruit

HOW BIG IS IT?
6 ft (1.8 m) tall, with a spread of about
4 ft (1.2 m).

WHERE'S IT FROM?
Areas of west Africa.

WHAT'S ITS NATURAL HABITAT?
Usually acidic soils with a pH of 4.5–5.8.

HOW'S IT POLLINATED?
By insects.

CAN I GROW IT AT HOME?
It's difficult—it needs humidity
and ideally an acid soil.

Okay, if any plant were my botanical party trick, it would be this one. For a few years now, I have been presenting "The Weird and Wacky Plant Show" around the world to (occasionally bewildered) audiences. About halfway through, I pull out a few dried-up red berries and invite random members of the audience to chew one. I'm soon serving them slices of lemon and handing them vinegar in shot glasses, ready for a flavor-bending sensation.

The miracle berry contains a glycoprotein called miraculin, which has the power to alter your ability to taste acidic foods correctly, as it binds itself to your taste buds. Consuming

THE FLAVOR-BENDING BERRY

Synsepalum dulcificum
LATIN NAME

acidic foods after chewing a miracle berry causes the miraculin in it to change your perception of sour to sweet and make acidic treats taste super sugary. So, those lemons will be transformed to sherbet, and the vinegar to something akin to cider. Plus, you can chew a miracle berry, then eat a strawberry, and have the sweetest candy you *never* wanted to eat, as the miraculin keeps cranking up that pH.

> ⌇⌇⌇⌇⌇⌇⌇⌇⌇⌇⌇⌇
>
> ## You can chew a miracle berry, then eat a strawberry, and have the sweetest candy you *never* wanted to eat.
>
> ⌇⌇⌇⌇⌇⌇⌇⌇⌇⌇⌇⌇

In West Africa, where the plant originates, the berries were always chewed to improve the flavor of corn bread and to sweeten palm wine. The effect lasts about an hour. Visitors to my show usually find it ruins— or rather improves—their own lunch. As for ill effects—as with anything, moderation is best, such as eating more than one bag of salt and vinegar chips.

In terms of a party trick, the possibilities of this fascinating berry are endless. Think beer, pickles, lemons, limes, improving the taste of your friends' dire culinary skills, and so on.

A godsend for diabetics?

Japanese researcher Kenzo Kurihara was the first to isolate miraculin in 1968, and in the 1970s researchers saw a bright future for this flavor-bending berry which, despite its low sugar content, had the potential to transform sour foods into sweet sensations. For the first time ever, you could reach sugary euphoria without even touching a

There are said to be tomatoes that have been bred to produce the same taste-modifying glycoprotein miraculin. They were originally designed to be a premeal snack to reduce the consumption of sugar.

doughnut—a godsend for diabetics. People could soon buy it in Japan in tablet form.

However, attempts to commercialize the berry were thwarted in the 1980s, as regulators would not pass it as a food stuff. Perhaps it was never commercialized because it would have meant the bottom falling out of the US doughnut market ... The mind boggles on that one.

But the miracle berry remains firmly on the European Union's list of novel foods (which means foods that have been traditionally eaten outside the EU) and care should be taken, because high consumption could increase the level of acid in the bloodstream—but one or two, eaten as a novelty, won't do any harm. "The Weird and Wacky Plant Show" can go on. Hurrah! Additionally, the leaf is used medicinally to help manage blood sugar levels, while the seed oil is used as a hair tonic.

Try miracle berries for yourself
Not being smug, but I have experienced miracle berries for myself when I had a live plant couriered to me from China a few years back. It didn't like our climate, though, and soon keeled over, so I resorted to freeze-dried berries. They are the next best thing and are the ones we use in my shows. You can easily find them online, so opt for those if you're in the market for changing up your bitter lemon to a fancy glass of Lambrusco.

Coco de mer, sea coconut, double coconut

THE LARGEST OF ALL SEEDS

HOW BIG IS IT?
The tree can reach 80 ft (25 m) or more.
The seed's diameter is 20 in (50 cm).

WHERE'S IT FROM?
Praslin and Curieuse islands,
in the Seychelles.

WHAT'S ITS NATURAL HABITAT?
Deep, well-drained soils and open,
exposed slopes.

HOW'S IT POLLINATED?
Possibly by lizards and geckos.

CAN I GROW IT AT HOME?
Nah, mate. Unless you live in the tropics;
although, acquiring a viable seed will be
more difficult than growing it.

The coco de mer's original botanical name, *Lodoicea callipyge*, comes from the Greek for "beautiful buttocks." So, there you go. Now that that's out of the way, we can discuss all the other attributes this fascinating tree has to offer.

To deal with the statistics first: *Lodoicea maldivica* holds the records for the largest seed in the world (20 in/50 cm in diameter); the longest-living flower head (it produces pollen for up to 10 years); and the world's longest cotyledon (the first leaf that any seed produces), at 13 ft (4 m).

But the most headline-worthy fact there is the dimensions of those gargantuan

Lodoicea maldivica

LATIN NAME

seeds. They're the same size as a rugby ball and, at 55 lb (25 kg), as heavy as the average 10-year-old child.

The reason they're so big is a result of a biological phenomenon known as "island gigantism," where the size of a species increases dramatically when it's isolated and away from competition. The great advantage of having huge seeds is, of course, that they tend to produce more vigorous seedlings. They have more power that way.

Until the coco de mer tree was found, many people thought that the giant seed came from a mythical tree that grew at the bottom of the ocean. Malay seamen had spotted them bobbing around in the waves. It was also thought that the seeds were distributed by water, but it was soon established that viable ones would be too heavy and that those washing up on shore were actually dead.

In coco de mer colonies, there are separate males and females, but determining the sex of the plants isn't as easy as turning a hamster upside-down in your hand. You have to wait until the flowers are produced, at between 11–45 years of age—a sort of puberty for the tree.

Determining the sex of the plants isn't as easy as turning a hamster upside-down in your hand.

Then, the butt-shaped seed develops on the female tree, while the male flowers transform into humongous, rather phallic, catkins. But how do the two get it together?

Enter the lizards and geckos
One pollination fantasy has it that on wild, stormy nights the male trees tear themselves up from the ground and become locked in a passionate embrace with the female trees. Of course, this is unlikely to be true.

Less fanciful, if not more believable, theories suggest that lizards and geckos could lend a helping hand. They flock to the flowers—along with white slugs—to gorge themselves on the tasty nectar and pollen.

With no obvious dispersal method for the heavy seed, the tree can't ever colonize much farther afield than its home islands or indeed its neighborhood. But *Lodoicea maldivica* rarely hurries anything. Once formed, seeds take six to seven years to mature on the plant and then two more to germinate.

Evolved to succeed
Over time, the coco de mer has found a way to create its very own habitat. Soil nutrients around the plant are some of the best for miles around. The 33-ft- (10-m-) long, fan-shaped leaves feature a clever network of pleats that funnel and flush pollen, bird feces, and leaf litter down to the soil below. Incredibly efficient. Furthermore, before shedding its old leaves, the tree sucks out every last drop of goodness from them. With less focus on the foliage, the seed is then getting all the attention. No wonder it gets to the size that it does.

The tree also takes care of nurturing its young. As seedlings germinate, they are raised in the shelter of their parent, in the most nutritious soil, thanks to the delivery of nutrients by that "foliar express service."

Tackling the risk of extinction
The coco de mer is, sadly, on the International Union for Conservation of Nature's Red List of Threatened Species, with habitat loss and exploitation of the very collectible seed putting it at risk of extinction. Things are looking brighter, though, with trade in the seed now being better controlled, and firebreaks being added to the landscape to protect colonies.

The catkins dangling down indicate that this is a male coco de mer tree.

HOW CURIOUS

Millions of years ago, when the Seychelles separated from India geographically, coco de mer trees lost the opportunity to spread, which is why they're unique to the islands.

COMMON NAMES

Titan arum, corpse flower

HOW BIG IS IT?
Up to 10 ft (3 m) tall, with a width of 5 ft (1.5 m).

WHERE'S IT FROM?
West Sumatra, in Indonesia.

WHAT'S ITS NATURAL HABITAT?
Rainforest openings on limestone hills.

HOW'S IT POLLINATED?
By carrion-eating beetles and flesh flies.

CAN I GROW IT AT HOME?
Now you've got me thinking ...

One of the wonders of the plant world, the titan arum has always been associated with outlandish rumors, from being pollinated by elephants, to the fear that cultivating one would put you at risk of being eaten by it. One rumor that's true, however, is the name—*amorphos* means "misshapen" in Ancient Greek and *phallus* does indeed mean "penis." It's the titanic misshapen-penis plant.

The titan arum isn't the largest flower in the world. That trophy sits with *Rafflesia arnoldii* (see page 146). But, it is the world's largest unbranched inflorescence (arrangement of flowers on a single stem). It's certainly debatable as to whether it's the smelliest plant, too, such is its stench.

A BLOOMING MARVEL

Amorphophallus titanum

LATIN NAME

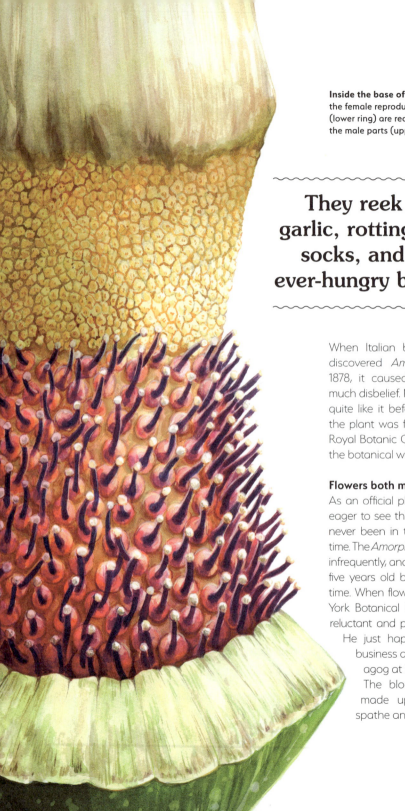

Inside the base of the flower spike, the female reproductive parts (lower ring) are ready a day before the male parts (upper ring).

They reek of cheese, garlic, rotting fish, sweaty socks, and feces. The ever-hungry beetles love it.

When Italian botanist Odoardo Beccari discovered *Amorphophallus titanum* in 1878, it caused a sensation and led to much disbelief. Nobody had seen anything quite like it before, and it was only when the plant was first cultivated—at London's Royal Botanic Gardens, Kew, in 1889—that the botanical world sat up and took notice.

Flowers both male and female

As an official plant geek, I've always been eager to see this plant in the flesh, but I've never been in the right place at the right time. The *Amorphophallus* blooms somewhat infrequently, and a plant must reach at least five years old before it flowers for the first time. When flowering occurred at the New York Botanical Garden in 2015, I sent my reluctant and plant-blind partner to see it. He just happened to be in town on business at the time, and even he was agog at its beauty (and its smell).

The bloom of the titan arum is made up of two basic parts—a spathe and a spadix, which sound as

In 2011, Roseville High School in California became the first school to bring a titan arum to bloom. Watch it for yourself on YouTube.

if they could be a New Romantic synth-pop duo. The spathe looks like an Elizabethan collar—it's a giant maroon petal, the color of dried blood, designed to be mistaken for a piece of meat by pollinators. The spadix resembles a slender baguette and is hollow, with two bands of flowers near the base. The upper deck is stuffed with male flowers, the lower with female flowers.

The females are ready to open first and, after a day, the males open, too. I think many heterosexual couples might dispute the fact that a female could ever be ready first, but nonetheless. This staggered opening is a mechanism to ensure pollination.

All in the timing

When it comes to getting pollinated, the odds are stacked against the titan arum. The window of opportunity is just 24–48 hours, the length of the flowering period. So another plant must be in flower nearby at exactly the same time—not easy when the titan arum has no specific flowering time (although it has been successfully self-pollinated in captivity a few times).

To raise its game, the plant can generate its own heat by a process known as thermogenesis, reaching 98.6°F (37°C) to help spread the fetid fragrance and attract pollinating insects. Its odors are stronger

after dark, when pollinators such as carrion beetles are at their most active. The fragrances aren't quite the type you'd find in your neighborhood pharmacy—they reek of cheese, garlic, rotting fish, sweaty socks, and feces. Ever-hungry beetles love it.

When it's all over, the bloom dies, but the show doesn't end there—after a dormancy of up to four months, a new leaf appears. Together, the leaf and stem can reach 20 ft (6 m)—taller than the bloom had ever been, and there's only one, standing in the forest like a showgirl. It acts like a giant solar collector, building up energy in the corm (a bulblike underground storage organ). Depending on how long this takes, a flower may or may not occur the following year—the blooms are unpredictable, with up to five years between them.

In 2013 the largest inflorescence in cultivation was celebrated at the Botanical Garden of the University of Bonn, in Germany. With a eye-popping length of 10½ ft (3.2 m) from the base of the corm to the tip of the inflorescence. The titan arum also holds the title of "largest-known corm," with the heaviest weighing 340 lb (153.9 kg) at the Royal Botanic Garden Edinburgh.

Flipping heck! It seems like the titan arum gives the greatest botanical show on earth. Next time, I want a ringside seat—with a clothespin on my nose just in case.

Great Basin bristlecone pine

NATURE'S OLDEST-LIVING TREE

HOW BIG IS IT?
16½–49 ft (5–15 m) tall, with a spread of 33 ft (10 m).

WHERE'S IT FROM?
The higher mountains of California, Nevada, and Utah.

WHAT'S ITS NATURAL HABITAT?
Large, open groups, where it often shares the habitat with other pines.

HOW'S IT POLLINATED?
By the wind.

CAN I GROW IT AT HOME?
Yes. It's actually not difficult to grow and would outlive you, your grandchildren, and many generations to come.

Gnarled, stunted, and somewhat waxy— no, it's not a Hollywood baddie I'm describing, it's the Great Basin bristlecone pine. This tree is officially the oldest-known, living, non-clonal organism on earth—in other words, the oldest tree grown from seed in the world. And, as this book goes to print, Methuselah, the oldest of them all, will be 4,855 years of age, according to *Guinness World Records*.

This means Methuselah's germination would have been in 2833 BCE—around the time the pyramids of Egypt were being built. Where did that first seed come from? We're into chicken-and-egg territory now.

Pinus longaeva

LATIN NAME

HOW CURIOUS

The habitat of the tree is so inhospitable and inaccessible that it's entirely possible there could be even older specimens out there.

The name bristlecone pine refers to the female cones, which have incurved prickles across their surface.

One thing is for sure, you won't ever be able to find Methuselah, as the exact coordinates of the oldest-known Great Basin bristlecone pine are, for its own protection, the secret of a select few. It is somewhere in the White Mountains of Inyo County, California. Scientists keep an eye on its condition, using an increment borer to extract a small, pencil-size core from the tree, which can then be analyzed. This process doesn't harm the tree, and the hole is covered over afterward with resin.

Edmund Schulman, Professor of Dendrochronology at the University of Arizona, was the first to recognize the species, and he and Tom Harlan, a researcher at the Laboratory of Tree-Ring Research in Tucson, Arizona, recorded this oldest specimen in the 1950s.

Secret of the trees' longevity

Will and persistence allow this incredible plant to reach such a ripe age. Branched, shallow roots maximize water absorption, while providing anchorage in hostile conditions. Waxy needles retain moisture and can photosynthesize for nearly 50 years, while the tree's bark is dark and resinous, making it super robust. Older trees can hang by a thread—sometimes only a narrow strip of tissue connects roots to the furthest branches.

But, the tree is perfectly happy. In fact, it does best on harsh terrain. The Great Basin bristlecone pine is a dominant species in high-elevation limestone soils, where few other plants want to grow. The gnarled shape of the plant is a symptom of this environment and I think the windswept stance gives it a somewhat theatrical appearance.

The tree is susceptible to fire, though. That resinous bark ignites swiftly and low-intensity burns even leave a scar. However, as a population, they are resilient and will quickly begin to repopulate, even if some are destroyed.

Aerodynamic aid to reproduction

But how does a tree in such a remote and hostile location reproduce? The answer is the wind. And not only does the wind pollinate the cones, it also distributes the seeds, which are easily airborne thanks to their aerodynamic wings. Clark's nutcrackers—birds that always opt for pine nuts—can play a part, too. When they pick the nuts, they bury them in the ground to store them for the winter. But, if they forget and don't go back for them, the seeds will grow into new plants.

The exact coordinates of the oldest-known Great Basin bristlecone pine are known only to a select few to protect it.

Yet older still

In a twist in the tale, dendrochronologist Tom Harlan actually measured a living specimen aged 5,063 years old. He kept the identity a secret until his dying day in 2013 and neither tree, nor confirmed data, could be found following his death. There are also 7,000-year-old trees known to researchers, but, alas, these specimens are long-dead.

Pinus longaeva was initially on the International Union for Conservation of Nature's Red List of Threatened Species. However, recent studies have shown sub-populations are no longer decreasing in size—so that's great news.

Asian watermeal, water eggs

THE SMALLEST FLOWERING PLANT

HOW BIG IS IT?
The size of a grain of sand.

WHERE'S IT FROM? Found in some areas of southern Asia.

WHAT'S ITS NATURAL HABITAT?
Quiet waterways, which it covers with a green, not unattractive goop.

HOW'S IT POLLINATED?
Botanists are unsure, but it could be by fish, birds, or the wind.

CAN I GROW IT AT HOME?
Some start-up companies are trying to find ways of growing it on the kitchen worktop, such is the interest in the plant's high protein content.

The world's smallest flowering plant is about the size of a single grain of sand. I said "sand" there, not rice. It's a plant so minuscule that with the naked eye you can't even tell when it's in flower. The tiny, transparent bodies without any stems, leaves, or roots are just $\frac{1}{250}$ in (0.1 mm) in diameter. It would take 5,000 of them to fill a thimble. As a consequence, the best images of *Wolffia globosa* are those taken under a microscope.

Commonly known as Asian watermeal, our tiny friend favors the warm, calm waters of Asia, where it carpets marshes, lakes, and ponds. It has also set up home in many

Wolffia globosa
LATIN NAME

The plant produces a flower with a visible stamen, anther, and stigma. Pollination isn't always necessary, however, as daughter plants keep appearing from the "budding pouch" at the side.

warmer parts of the United States, but refuses to grow easily in shade, which could be a relief, considering how invasive it can be.

Multiplying by seed and buds

During the summer, a microscopic bloom is cradled in the indent on the top of the plant. Exactly how the flower gets pollinated is still unexplored. Botanists think that it could be done by fish, birds, or even strong winds. Either way, a tiny, tiny fruit can indeed be produced—and would you believe that *Wolffia globosa* also holds the award for the world's smallest fruit, which contains just one tiny seed.

The plant doesn't really waste time waiting on pollination, though, since it can also reproduce asexually and exponentially by budding. Splitting and dividing every 36 hours to produce "daughter plants" in

quick succession—like something out of a horror film—it grows rapidly. A population of a nonillion (that's one followed by a whopping 30 zeros) plants can appear within just a few months. The daughter plants are produced within a small cavity at the base of the plant called the "budding pouch."

Later in the year, resting buds may sink to the bottom of the water in cooler countries to stay frost free, rising again to the surface in spring, in synchronized-swimmer fashion.

Tasty and nutritious?

It is possible to grow Asian watermeal at home, however careful water management is needed. Plants also need warmth to grow, as I found when I ordered a vial of *Wolffia* from the Czech Republic a few years ago and set it up in a jar at home.

The plant is 165,000 times smaller than the tallest eucalyptus trees, which is pretty cool.

It sadly died within a week in my cold home. One pesky characteristic of growing the plants for home consumption is that they soak up trace elements in water, including chemicals and pollution—great for water purification, but not for your belly.

Wolffia is therefore toxic when fresh, but drying or boiling it transforms it into an Asian foodstuff that's surprisingly rich in protein (20 percent, no less); carbohydrates; and fat; as well as vitamins A, B2, B6, and C. The name "watermeal" comes from the appearance of the small, mealy particles. So, if you're up for it, the dried stuff is edible and extremely tasty, with a sweet cabbage flavor—I like to think of it as "green caviar."

Most often used in Thai cuisine, where it is called *khainam* (meaning "water eggs"), it's a plausible food source that could be more space efficient than soy beans—watch this space. I'll definitely be hunting down a bowl of Asian watermeal on my next visit to Thailand. In some regions, it's used as an omelet filling. I rather fancy trying that.

> A population of a nonillion (that's one followed by a whopping 30 zeros) can appear within just a few months.

COMMON NAME
Corpse flower

HOW BIG IS IT?
The flowers measure 2–3 ft
(60–90 cm) across, usually appearing
close to the ground.

WHERE'S IT FROM?
The rainforests of Sumatra
and Borneo.

WHAT'S ITS NATURAL HABITAT?
Inside rainforest vines. It only emerges
when it flowers.

HOW'S IT POLLINATED?
By flies and beetles attracted by the
"nasty" smell.

CAN I GROW IT AT HOME?
Absolutely impossible. Sorry.

At 2–3 ft (60–90 cm) across, the *Rafflesia arnoldii* can lay claim to being the world's largest flower. But, despite its swelling physical presence, it's actually quite shy and elusive. It spends most of its life concealed within a host plant, the *Tetrastigma* vine, "ghosting" itself inside until it's ready to show off to the world. It's parasitic, with no chlorophyll of its own, and takes all the water and nutrients it needs from the host.

When *Rafflesia* is ready to bloom, it spends almost a year producing a flower bud, which develops outside the vine and resembles a big,

THE WORLD'S LARGEST FLOWER

Rafflesia arnoldii

LATIN NAME

The flowers appear within *Tetrastigma* vines and bloom for only a few days before dying.

blood-and-flesh colored cabbage. When the flower bursts open, the full glory is revealed. It's hard to believe it's real—the color, size, and texture look otherworldly.

The sight of the flower isn't the only thing that gets people talking—the smell of it is akin to rotting flesh, which is why the *Rafflesia* is sometimes called the corpse flower. This stench has evolved to attract pollinating flies and beetles to the "over-ripe meat buffet." The flower even goes a stage further—it has a network of appendages inside that look like hairs and fur, to carry on tricking the pollinators into thinking it's just a piece of innocent meat and not a flirty flower desperate to be pollinated.

The perfect parasitic marriage

It was once thought that elephants pollinated the flowers. In fact, it's the parasitic insects that live on the elephants.

Pollination isn't a straightforward process, as the insects, such as flies and beetles, must first visit a male flower and then a female flower in close succession. It's trickier than getting your groceries packed into the correct bags. Given how rarely blooms appear, things really aren't easy. Then, distribution of the 1,000 or so seeds is also kind of tricky. The flower relies mainly on tree shrews to spread them by eating and pooping. But, to survive, the seeds must end up near a *Tetrastigma* vine, which is the only plant that can host the *Rafflesia*. Luckily, vines can host multiple flowers, but just thinking about the whole precarious process makes me nervous.

A personal sighting

Despite the flower's size, your chances of seeing one in the wild are slim. Flowers only last a handful of days and deforestation in the areas it grows in means plant numbers are dwindling. Residents on Sumatra who find the plants popping up on their property are encouraged to preserve them. Some even charge visitors to see this wonder of nature, but that's a double-edged sword—while conservation of *Rafflesia* is intended to attract visitors, the tourists can damage their habitat. In a bid to preserve the *Rafflesia* family, environmentalists are trying to recreate their exact growing conditions.

I was very fortunate to go to Borneo a few years ago and was taken to see *Rafflesia keithii* in the wild. This species is not so large, but no less spectacular. I stood agog for at least four minutes and actually asked if it was a ceramic stand-in for the plant, such was my delirium. I guess I was just super lucky that day. Anyway, it's another big tick off my horticultural bucket list.

It was once thought that elephants pollinated the flowers. In fact, it's the parasitic insects that live on the elephants.

The largest bloom ever recorded was 44 in (111 cm) in diameter and weighed 24 lb (11 kg)—that's more than a well-fed toddler.

COMMON NAME
Wollemi pine

HOW BIG IS IT?
80–130 ft (25–40 m) tall, but branching can be complex, so spread is only ever estimated.

WHERE'S IT FROM?
Wollemi National Park, in New South Wales, Australia.

WHAT'S ITS NATURAL HABITAT?
Temperate rainforest within a series of remote, narrow, steep, sandstone gorges.

HOW'S IT POLLINATED?
By the wind, most likely.

CAN I GROW IT AT HOME?
Yes, it's a lovely tree for your yard or as a Christmas tree indoors.

When I started writing this book, I wanted to include some new plants. But, alas, they don't always have a wildly exciting story to them, unless you go all "X Factor" and start including side narratives, such as the owner's heartbreak over a spaniel with an unexplained rash. So you can imagine my glee when I realized I could include the Wollemi pine from Australia, a tree that has only recently been discovered and one with its fair share of drama, peril, and intrigue.

For 40 million years, the genus the Wollemi pine belongs to was AWOL. Fossil records were the only proof of this fabled group of trees, which disappeared as a

THE LUCKIEST TREE

Wollemia nobilis
LATIN NAME

No sooner had the Wollemi pines been discovered, however, than they were at risk again.

result of prehistoric climate change when the continent of Australia drifted and dried out.

Then, in September 1994, David Noble, Michael Casteleyn, and Tony Zimmerman were out looking for canyons. David Noble worked for the National Parks and Wildlife Services (NPWS) and had good botanical knowledge. To his astonishment, he spotted a grove of what he thought were the extinct pines. Collecting a dead piece of branch, he took it to Wyn Jones, a colleague at NPWS, and Jan Allen, from the Royal Botanic Garden Sydney, who were able to confirm the identity of the tree.

At this point, the tree had not previously been given any sort of official name. It was only later that it became known as the Wollemi pine, named after the national park in which it was found, although it isn't actually a pine, but a relation of the monkey puzzle tree (any conifer-like tree is often dubbed a pine).

Secret operation

Given how plentiful the trees were said to be in eastern Australia more than 40 million years ago, this was an amazing discovery. The damp, steep-sided gorge where they were found had somehow managed to avoid fires and destruction over the centuries.

To protect the trees, the finding was kept secret for months while the NPWS and the Royal Botanic Garden Sydney sent helicopters to collect cones.

Down at ground level, the researchers found counting the trees a challenge, as Wollemi growth is keen on branching out, but records show there were around 100. Each one was numbered and labeled, so when propagation material was gathered, it was known which plant it had been taken from. Each of the wild plants has since been replicated in the collections at the Royal Botanic Garden Sydney.

On investigation, it's thought the trees had experienced a phenomenon known as "genetic bottlenecking," with their total isolation and constant inbreeding leading to a loss of genetic diversity.

Ups and downs in Australia

No sooner had the Wollemi pines been discovered, however, than they were at risk again. Despite their location remaining a secret, and visitor equipment being sprayed with methanol to sterilize it, in November 2005, some of the wild trees showed signs of Phytophthora. It's not known how this potentially fatal water mold reached the remote canyon, but, by a whisker, the colony survived after careful management.

As if that wasn't drama enough, there was more to come. During the bushfires of 2019–2020, the whole colony was threatened by some of the largest wildfires in modern

Australian history, being saved in the nick of time by a dedicated team of firefighters from the NPWS, who dropped retardant from helicopters and set up additional irrigation around the plants. To keep the location secret, the survival rescue wasn't reported in the news at the time. In fact, the location of the gorge is still on a "need-to-know" basis.

The more the merrier
Unsurprisingly, Wollemi pines are listed as critically endangered on the International Union for Conservation of Nature's Red List of Threatened Species, and are protected in Australia. Thankfully, there are initiatives pushing for commercial production of the pine: the more trees that are grown, the more the genetics can be preserved. Its potential as a Christmas tree has also been spotted, so soon you might see it hanging with baubles. That would be the perfect way to celebrate the salvage operation of this important piece of history.

HOW CURIOUS
Wollemi pines grow better in the UK than in Australia—a sign that the climate was much wetter there all those millions of years ago.

There is a huge drive to plant Wollemi pine saplings to mitigate the risk of this rare tree becoming extinct.

X-RATED PLANTS

05

Time to put your inhibitions to one side. Plants are sexual—get over it. And in this chapter, it isn't just their names that sound rude—their appearance and their behavior can sometimes be pretty saucy, too. But, whether they make you blush or trigger a snigger, it's all in the name of botany.

COMMON NAMES

Naked man orchid, Italian orchid

THE COLONY OF NUDISTS

HOW BIG IS IT?
20 in (50 cm) tall, with a spread
of 10 in (25 cm).

WHERE'S IT FROM?
Southeastern and southwestern Europe,
western Asia, and North Africa.

WHAT'S ITS NATURAL HABITAT?
Abandoned farmland, scrubland, and
macchia (dense evergreen shrubland).

HOW'S IT POLLINATED?
We're not completely sure. Probably
by insects, the experts think.

CAN I GROW IT AT HOME?
Difficult. It has quite specific soil and
fungus requirements.

**I'm rather accidentally becoming known
for rude plants. I'm not really that smutty,**
honest. It's just that I find them fascinating.
While certain plants can look like wildlife,
birds, or even genitalia, they are seldom
designed that way. Their primary purpose
is to attract pollinators, not to make
horticultural students giggle and guffaw.

One such plant is *Orchis italica*. The
labellum (lobed lip of the central petal) of
each floret undeniably resembles the shape
of a naked man. You can detect the arms,
legs, torso—and the other thing. He's even
wearing a hoodie (although in actual fact
it's the bloom's outer tepal) and he seems

Orchis italica
LATIN NAME

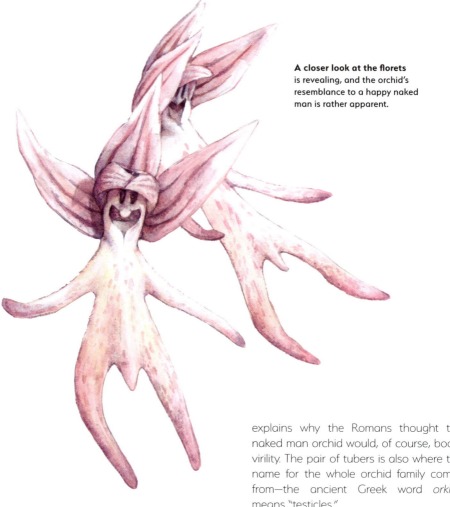

A closer look at the florets is revealing, and the orchid's resemblance to a happy naked man is rather apparent.

to be smiling, thanks to some neat little purple markings on the inner tepal. Each stem can host up to 40 florets, very often more, not unlike a working men's social club in the 1960s.

Testicle-shaped tubers

Beneath the ground, *Orchis italica* grows as a pair of tubers that resemble testicles. In times gone by, it was believed that, if a plant looked like an organ of the human body, then it could assist that organ, which explains why the Romans thought the naked man orchid would, of course, boost virility. The pair of tubers is also where the name for the whole orchid family comes from—the ancient Greek word *orkhis* means "testicles."

The tubers of *Orchis italica* can be dried and ground down into an arrowroot-like powder known as salep. One ounce of salep is so filling it's said to be able to keep someone going for a whole day. Most popular in Turkey, salep is made into a delicious bread, a hot beverage, or even ice cream. Sadly, salep is enjoyed so much that these orchid plants now have threatened status, as tubers have been harvested from wild plants over the years with reckless abandon.

In its natural habitat, *Orchis italica* has a symbiotic relationship with a fungus in the soil, much like *Caleana major*, another fabulous orchid with intriguing blooms (see page 54). The fungus ensures the orchid obtains the right nutrients and helps it compete with other plants. In return, the orchid provides the fungus with sugars.

Replicating these conditions with the exact fungus is tricky—and the plant is sensitive to any added fertilizers or fungicides. So, sadly, it's hard to grow at home.

A sight for sore eyes

Despite its exacting needs when growing in the wild, the naked man orchid can very occasionally pop up uninvited in gardens. (And the week after that happens, you'll win the lottery ....)

Orchis italica is the most familiar of the Mediterranean orchids and is arguably the most abundant. I am told that nothing can prepare you for the sight of colonies of the naked man orchid across hillsides during the spring. I'd be down on my knees inspecting those blooms up close before you could say "botanical treasure." Give me that Greek vacation now.

HOW CURIOUS

The flowers are said to have the fragrance of freshly mowed hay. Bottle me up some of that, please.

Beneath the ground, *Orchis italica* grows as a pair of tubers that resemble testicles.

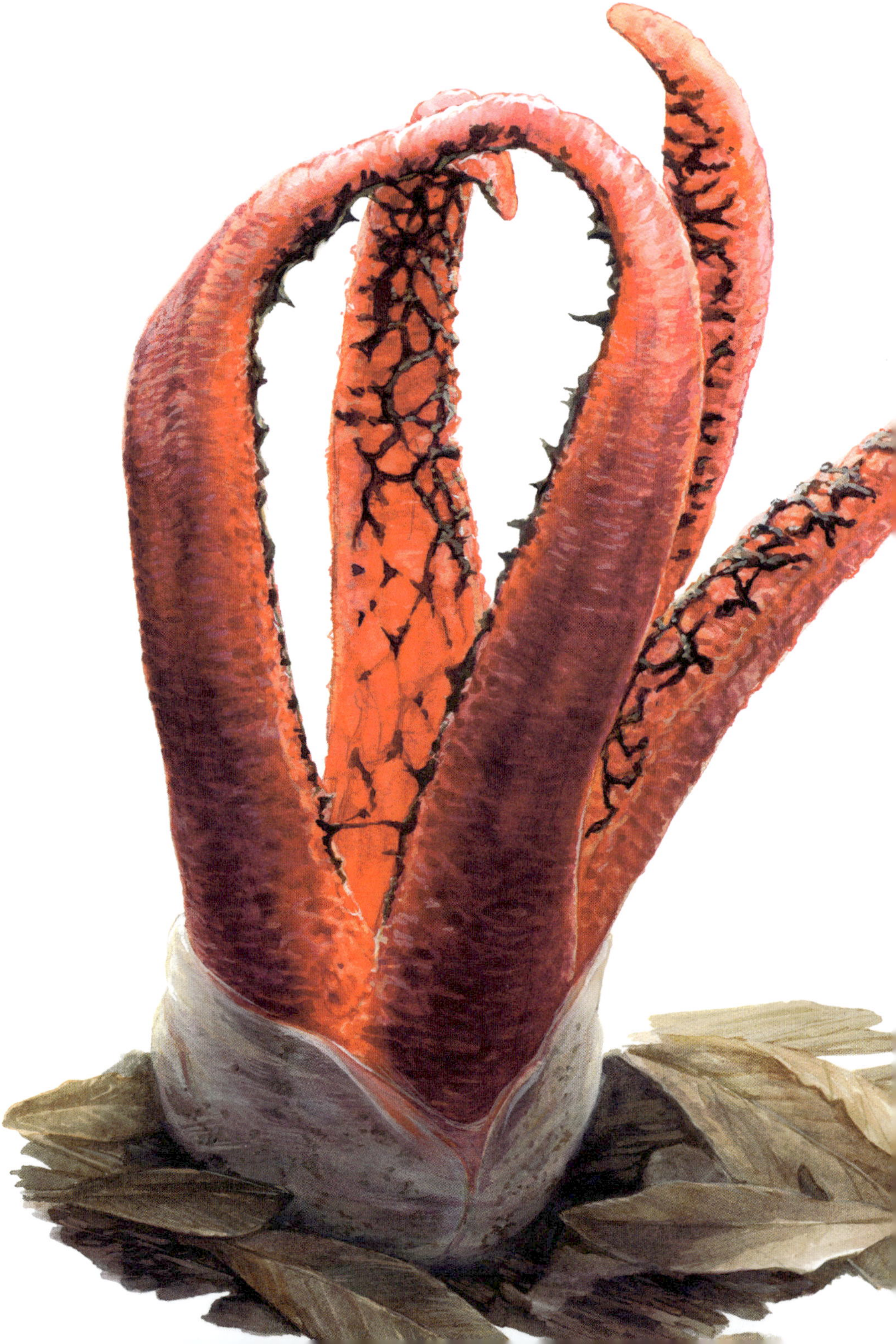

Octopus stinkhorn, devil's fingers

THE FUNGUS THAT COULD FLIP YOU OFF

HOW BIG IS IT?
4 in (10 cm) tall, with a body width of 8 in (20 cm), and tentacles 2¾ in (7 cm) long.

WHERE'S IT FROM?
Australia and New Zealand.

WHAT'S ITS NATURAL HABITAT?
Leaf litter beneath trees and shrubs, and sometimes in bark mulch in parks and gardens.

HOW DOES IT REPRODUCE?
Insects and slugs spread the spore-filled mass.

CAN I GROW IT AT HOME?
Quite tricky, so best leave it to nature.

I've never been a fan of horror films—I simply don't see the point in scaring myself for the sake of it. So, as I recently hunted around London's Royal Botanic Gardens, Kew, after a reported sighting of devil's fingers, I had my heart in my mouth. Alas, I couldn't find it, despite the reliable-sounding tip-off on Instagram. So, to date, I have only seen this incredible fungus in books and on the old internet.

To describe *Clathrus archeri*, let's set the scene: the first thing you might notice as you're walking past an unassuming pile of fall leaves is a slimy white "egg," not unlike a golf ball, emerging from the ground. This

Clathrus archeri

LATIN NAME

is actually the fruiting body of the fungus, and what happens next is the stuff of low-budget horror films. Inside, there are tentacles forming, and in time, through the white outer membrane, you can see their pinky mass squirming around.

Quite soon after, between four and seven elongated red "arms" appear, linked at the top like a cage (the name *clathrus* means "barred" in Latin). The arms are the shade of fresh meat and are coated in gleba—a gooey, putrid-smelling mass of millions of spores—which takes on the form of dark, clotted blood.

These arms later reach out, beckoning forth snack-hungry insects and even slugs to devour the slime on its creepy tentacles. As the arms spread, the mass becomes more and more attractive to wildlife. Some of the insects slurp up the goo and eject the spores in their poop; others unintentionally carry it away on their feet and undersides. Clever work, Mr. Fungus—rather than relying on wind and rain, you have an army of forest creatures at your command.

> **Some of the insects slurp up the goo and eject the spores in their poop; others unintentionally carry it away on their feet and undersides.**

This method of spore distribution allows the fungus to spread far and wide. As long as the spores land in a space with moisture and food, they can germinate there when the time is right and temperatures are warmer and consistent enough. As they grow, they'll produce hyphae—small underground threads that gather food and nutrition—creating a network of them under our feet. The fingers are merely the "fruits" of the fungus.

HOW CURIOUS

It's thought that the fungus spores arrived in Europe as "stowaways" on boats bringing war supplies from Australia during World War I.

Fancy a fright night?

Although devil's fingers is a rare find in the UK, it can sometimes be spotted in Cornwall and the New Forest. The fungus is saprobic, which means it lives on organic rich matter and needs very little oxygen, so look for it in the leaf litter beneath trees, and among the bark mulch in your local park. From June to September is the best time.

Clathrus archeri isn't toxic. In fact, the eggs are edible, if you fancy giving them a try. But take care when harvesting any type of wild fungi or fruit, and consult an expert to ensure correct identification—look at me going all caring on you!

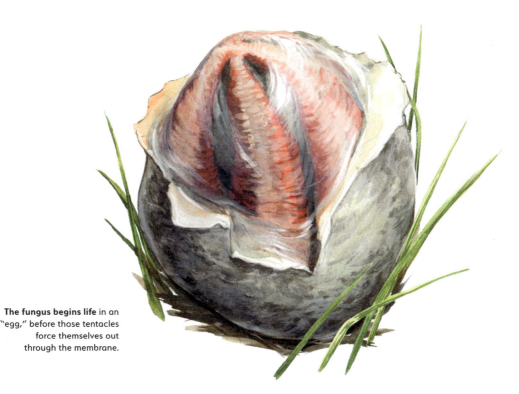

The fungus begins life in an "egg," before those tentacles force themselves out through the membrane.

Butterfly pea, Asian pigeonwings

THE
VULVA VINE

HOW BIG IS IT?
A scrambling vine, this can grow and spread around 10 ft (3 m).

WHERE'S IT FROM?
Thought to be mainly from tropical Asia and Africa, although it has since made itself at home in many other locations.

WHAT'S ITS NATURAL HABITAT?
Grassland and open woodland, generally seen clambering over other vegetation.

HOW'S IT POLLINATED?
By insects.

CAN I GROW IT AT HOME?
Yes. It's a fun summer climber for the patio.

There's a lot more to this plant than meets the eye. I'm so glad I found it. With applications ranging from coloring gin and Thai sticky rice, to its use as an ingredient in beauty products, *Clitoria* is a mightily hardworking plant. It can even help improve the soil it grows in.

But, before we go any further, let's get the obvious out of the way. The Latin name "clitoria" clearly alludes to the flower's appearance, and in most countries even common names refer to the resemblance it bears to a part of the female genitalia. Despite protests by a minority of very conservative botanists, who saw the name

Clitoria ternatea

LATIN NAME

The plant is perfect as a decorative wall climber, with its wisteria-like appearance.

as vulgar and immoral, less explicit alternatives have never caught on.

Although it's a somewhat tropical vine, I grew one in a pot once and, despite the fact that it didn't flower until the last few days of summer, it was worth it, for its mystical wisteria-like vibe. And, actually, it flowered sooner than a wisteria plant would have done anyway. Wisteria are known for taking almost 10 years to flower from planting, and even then they only do so with the correct pruning. With *Clitoria*, you get the same ethereal blue-colored flowers, as well as the wisteria-like foliage. It needs support, though: it almost can't be bothered to climb, usually wanting to clamber over its neighbors, even competing with weeds.

To dye for

If you're in a region where you can go "tropical foraging," grab a rustic basket and pick a few of these azure beauties, because it's when you get *Clitoria* into the kitchen that things get really interesting. You can include the flowers in rice pudding to make it tropical blue. In Thailand, they are used to color sticky rice and served with golden mango in a wicked color combo—the best street food, and loved by locals.

Or you can make a cup of vivid-blue herbal tea from the dried flower heads (easily available online). Just steep them in boiling water for a few moments and, if there's any left over, make some blue ice-cubes with it.

The color comes from a compound called anthocyanin, which is also what makes blueberries blue. Curiously, when you add tonic water to gin colored with *Clitoria* flowers, the whole drink ombrés from blue to pink right before your eyes, thanks to a change in pH—yet there's no change in the flavor.

Increased soil fertility

Like many plants in the legume (pea and bean) family, *Clitoria ternatea* has a symbiotic relationship with certain soil bacteria and can form nodules on its roots that fix atmospheric nitrogen into the soil. This is a win-win—you don't just get the beauty of the *Clitoria* plant, you're improving the soil, too.

I've also seen notes about the plants being used as an eco-friendly insecticide. That's well worth exploring, I'd say. Therapeutically, *Clitoria* is popular in Ayurvedic and Chinese medicine. I think they said it's good for memory ....

> **Therapeutically, *Clitoria* is popular in Ayurvedic and Chinese medicine. I think they said it's good for memory ...**

HOW CURIOUS

Clitoria is considered a holy flower in India, and is used daily in Hindu puja rituals.

COMMON NAMES

Hairy balls, balloon plant

HOW BIG IS IT?
Up to 8 ft (2.5 m) tall, with a spread of 3 ft (1 m).

WHERE'S IT FROM?
Southeast Africa, from South Africa to Mozambique, but now widely naturalized elsewhere.

WHAT'S ITS NATURAL HABITAT?
Roadside banks and grassland.

HOW'S IT POLLINATED?
By vespid wasps and hornets.

CAN I GROW IT AT HOME?
Yes, and won't it be worth it to shock the neighbors? It does need a good warm summer, though.

When I first started seeing the smutty side to plants, if somebody looked offended, I'd exclaim, "it's not rude, it's botany." This motto has served me well. You see, I first came across this mega-cool plant when I was visiting the legendary seed company K Sahin, Zaden BV, in the Netherlands. The trials team there used to experiment with lots of quirky annuals and this one was simply introduced as *Gomphocarpus* "hairy balls." Little space for nuance there.

So, let's not beat about the bush—the inflated seed pods of *Gomphocarpus physocarpus* do indeed look like unshaved

THE PLANT WITH BALLS

Gomphocarpus physocarpus

LATIN NAME

testicles. In comparison, the flowers that precede them are absolutely angelic—white, waxy, and skirted, with the sweetest scent of candy.

A sting in the tale

An important role in the pollination of *Gomphocarpus* flowers is played by the humble and often-misunderstood wasp. Vespid wasps, surprisingly, don't actually want to sting you as you drink your summer beer—they'd rather be sipping from the cup of nectar that flowers like *Gomphocarpus* provide. The blooms are randomly visited by other insects, too, but it is the hornets and vespid wasps that commit to completing the task of pollination.

> ## You might see a few "hairy balls" in your local florist, as the plant has found favor as a cut stem when it's in flower.

Each flower has a complex shape, with dark slits that the insects often pop a leg inside to help stabilize themselves. Little do they know that inside the slits is a bath of sticky pollen, grains of which get stuck to their legs. Once the wasp finally pulls its leg free, it travels to the next flower, where it gets trapped all over again. But, in the process, it has transported pollen from one plant to another.

Gomphocarpus physocarpus is self-incompatible, which means it will set seed only when pollinated by another plant, albeit of the same species and type. Regardless, it is fairly successful in reproducing and was classified as a weed in Australia as

When the pod is ripe, it explodes and releases the seeds. Their tufts allow them to be carried by the wind to places they can germinate once they've landed.

Although perhaps not what this plant is most known for, hairy balls also produces delicate pink-white flowers, as well as those iconic seed pods.

HOW CURIOUS

The plant's name *Physocarpus* comes from *physa*, the Greek for "bladder," and *karpos*, meaning "fruit." The rest you can figure out for yourself …

early as 1868, which is pretty serious considering that the plant was only discovered that same year. It's also toxic—something that the caterpillars of the monarch butterfly are seemingly immune to, and they have quite a time of it feasting on the plants.

Keep an eye out for hairy balls

These days, you might see a few "hairy balls" in your local florist, as the plant has found favor as a cut stem when it's at the pod stage. If you're ever in North America, you may also spot hairy balls used as a centerpiece annual in flowerbeds.

Given the kink status of this becoming plant, you can imagine my glee when I found a patch of hairy balls 8 ft (2.5 m) high at the Toronto Botanical Garden in Canada a few years back. And, yes, I posed for a photo.

COMMON NAMES

Exploding cucumber, squirting cucumber

HOW BIG IS IT? 3 ft (1 m) tall, with a spread of 2–3 ft (60–90 cm).

WHERE'S IT FROM? Southern Europe, North Africa, and Western Asia.

WHAT'S ITS NATURAL HABITAT? Uncultivated areas, such as roadsides, where it can get invasive.

HOW'S IT POLLINATED? Usually by bees.

CAN I GROW IT AT HOME? Yes, and you can do the party trick with it. But make sure you wash your hands afterward, as exploding cucumbers are highly toxic.

Looking back at my vacation photos from a previous summer, I have many blurred ones from the Greek mountain village of Arachova. The pictures tell the story of me trying to tease an *Ecballium* into ejaculating its seed-filled liquid across the Greek hillside. There are even videos showing my attempts, plus my scream when it actually happens. You see, the force of an exploding cucumber distributing its seeds is akin to the sting from a snapped elastic band and it's very surprising.

The *Ecballium* is the king of rapid plant movements, up there with the waterwheel

THE SHOOTING FRUIT

Ecballium elaterium

LATIN NAME

plant (see page 12) and the telegraph plant (see page 118). As the cucumber explodes, its seeds shoot through the air at speeds of up to 60mph (95kph), landing as much as 33ft (10m) away. By distributing its seeds with such force, the plant can easily colonize the surrounding area. This it does very successfully, which is why it is classified as invasive in many regions. Thankfully, the seeds I managed to eject so skillfully only ended up staining my best summer shorts.

The pressure's on

My foggy brain shall now attempt to give you a rough explanation of what makes the fruits explode. It's a phenomenon known as "irreversible turgor movement" and it's all to do with pressure.

As the cucumber ripens, it bulges with around 10–20 seeds inside it, all wrapped up in a viscous mucilaginous liquid. The cucumber is surrounded by a *pericarp* (the ripened wall of the ovary), which is multi-layered and elastic. As the seeds within the fruit start to grow in size, the pressure on the pericarp increases, but it can resist the pressure to begin with, thanks to its elasticity. However, when the seeds have become ripe and so large that the pressure is high enough for a "critical value" to be surpassed, the less resistant tissue at the base of the fruit ruptures, releasing the cucumber from the plant. And that's when the catapulting fun begins.

As the cucumber bursts away from the stem, the cells between the seeds rupture, and they spurt out of the fruit at a ballistically ideal angle of 40–60°, causing the fruit to shoot through the air.

It's this that I was putting to the test on that roadside in Greece. As I was caressing the bottom of the cucumbers, only the yellow ones were exploding, because they were the ripest, with the most well-developed seeds inside, which were exerting the greatest pressure on the connecting tissue.

> ## As the cucumber explodes, its seeds shoot through the air at speeds of up to 60 mph (95 kph), landing as much as 33 ft (10 m) away.

A very cool cucumber

Exploding cucumber plants are usually found straggling along the roadside, as they have no tendrils to climb. Some people like to grow them as ornamental plants—if only so they can perform the best party trick of any plant.

While you may not easily mistake a squirting cucumber for a classic cucumber, I would also recommend that you *really*

Although it's part of the melon family, eating an *Ecballium* isn't advised, and it certainly wouldn't taste like honeydew.

don't, as *Ecballium elaterium* is highly toxic. In folk medicine, it's often used as a purgative, which means that it will give you "a good clear-out."

If you'd like to watch an *Ecballium* explode for yourself, why not go on YouTube. But I would advise you to delete your internet history afterward ...

***Ecballium elaterium* have small,** yellow flowers. From left to right here, are the male and female flowers in bloom, and the pollinated female flower, bearing the notorious fruit.

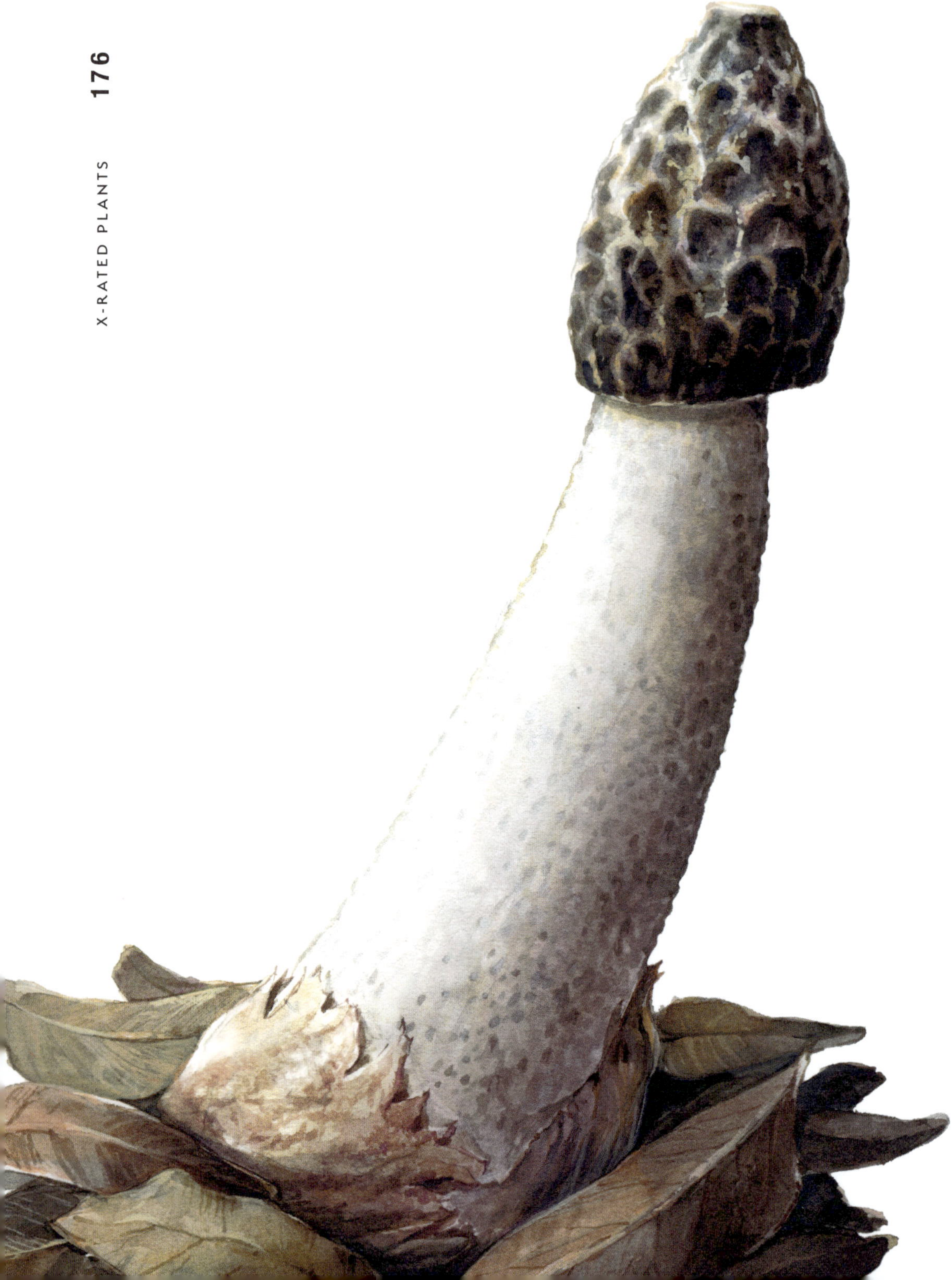

Common stinkhorn

A BRAZEN EXHIBITIONIST

HOW BIG IS IT?
The fruiting body is 10 in (25 cm) tall, and 2 in (5 cm) wide.

WHERE'S IT FROM?
Europe and North Africa, but often found farther afield, too.

WHAT'S ITS NATURAL HABITAT?
Wood debris, forest floors, and mulched gardens.

HOW DOES IT REPRODUCE?
By flies' legs and sometimes their poop.

CAN I GROW IT AT HOME?
It will probably make that decision for you.

Let's break down that botanical name, first of all. Swedish botanist Carl Linnaeus knew exactly what he was doing back in 1753 when he gave this saucy-looking fungus the name *Phallus impudicus*. As the word suggests, *phallus* means "penis" in Greek. But, it's *impudicus* that really tips the wink, as it means "shameless" or "immodest" in Latin. Turn the page now if your smut-alarm is already letting off cautionary bleeps.

The first thing you'll see is a white "egg" or two. At this stage, the fungus is referred to as a witch's egg. Soon enough, an erect fruiting body pushes through in shocking fashion. Tall; white; and with a conical, slimy head, it grows at a speed that's been

Phallus impudicus

> ～～～～～～～～～～～
> ## You'll likely catch a whiff of rotting meat before you see the fungus itself.
> ～～～～～～～～～～～

measured at 4 in (10 cm) an hour. It's believed it can even push its way through asphalt.

It's this slimy head that contains the spores, in the gleba (spore-bearing liquid) that covers it. Insects are attracted by the constant smell of rotting meat, ensuring that the tip of *Phallus impudicus* stays as busy as a small-town club on New Year's Eve.

The vibrations of this busy event cause the gleba to ooze a sugary liquid mixed with spores, flooding the "dance floor" with

The common stinkhorn grows by "hatching" out of an "egg" at an alarmingly fast rate. The fungus can push out and form within a single day.

At the "egg" stage, the fungus is often eaten in France and Germany and is said to taste of hazelnuts.

a gooey mass that quickly gets slurped up by the insects. When they later answer the call of nature, they unknowingly start a new colony of this *Phallus* species nearby. Their dancing feet also take a few spores on to their next destination, increasing the infinite spread of this fungus as they do so.

Dead useful

Phallus impudicus should be welcomed in your garden, not feared. This fungus is "saprobic" which means it feeds on dead and decomposing plant material, and fulfills a very important role as a decomposer in any ecosystem they find themselves in. This particular fungus also goes to the top of the class for its mycorrhizal associations with nearby trees.

In a mycorrhizal relationship, both parties benefit. Mycorrhizal fungi improve the nutrient status of their host plants, influencing how they absorb nutrients and improving their water uptake and disease resistance. In exchange, the host plant is exploited for fungal growth and reproduction purposes.

Follow your nose

If you ever find yourself up for a spot of fungi hunting any time between June and November, make sure you take some smelling salts in your backpack. Sniff your way into the forest to find this gem, since you'll likely catch a whiff of rotting meat before you see the fungus itself. Then, you'll be grasping those salts like a good 'un. If you're searching for tasty morels, which look remarkably similar, you won't get the two mixed up, as the stench of the common stinkhorn will totally give it away.

In Montenegro, handfuls of the fungi were often smeared on bulls' necks to make them stronger during bullfights. They were also fed to the young bulls, and eaten by people across Europe, on the basis of their being an aphrodisiac. The Victorians, on the other hand, found them embarrassing. That's perhaps all you need to know about the European passion polls.

BIBLIOGRAPHY

Simon Barnes, *The Green Planet*, BBC Books, 2022.
Jonathan Drori, *Around the World in 80 Plants*, Laurence King, 2021.
Ambra Edwards, *The Plant-Hunter's Atlas*, Greenfinch, 2021.
Don Ellison, *Cultivated Plants of the World*, New Holland Publishers, 1999.
James Hoffmann, *The World Atlas of Coffee*, Mitchell Beazley, 2018.
Stefano Mancuso, *The Revolutionary Genius of Plants*, 2018.
Dan Saladino, *Eating to Extinction*, Jonathan Cape, 2021.

Australian National Botanic Gardens Centre for Australian National
    Biodiversity Research, www.anbg.gov.au
Beekenkamp Plants BV, www.beekenkamp.nl
Cambridge University Botanic Garden, www.botanic.cam.ac.uk
Carnivorous Plant Resource, www.carnivorousplantresource.com
The Eden Project, www.edenproject.com
First Nature, www.first-nature.com
Game and Wildlife Conservation Trust, www.gwct.org.uk
Hortus Botanicus, Leidon, www.hortusleiden.nl
In Defense of Plants, www.indefenseofplants.com
Internet Orchid Species Photo Encyclopedia, www.orchidspecies.com
The Invasive Species Compendium, www.cabi.org
National Park Service, www.nps.gov
The National Wildlife Federation, www.nwf.org
Natural History Museum, www.nhm.ac.uk
Niigata Botanical Garden, https://.botanical.greenery-niigata.or.jp
Orchids of Britain and Europe, www.orchidsofbritainandeurope.co.uk
Plants for a Future, www.pfaf.org
RHS Garden Wisley, www.rhs.org.uk/gardens/wisley
Royal Botanic Garden Edinburgh, www.rbge.org.uk
Royal Botanic Gardens Kew, www.kew.org; https://powo.science.kew.org
Toronto Botanical Garden, www.torontobotanicalgardens.ca
Useful Tropical Plants Database, https://tropical.theferns.info
The Wildlife Trusts, www.wildlifetrusts.org
Woodland Trust, www.woodlandtrust.org.uk
World of Flowering Plants, www.worldoffloweringplants.com

Plants Behaving Badly

Armstrong, W.P., "Unforgettable Acacias," *Zooznooz*, 1998. Accessed through Palomar College, www.palomar.edu

BioNET-EAFRINET, Invasive Plants, www.lucidcentral.org/keys/v3/eafrinet/plants.htm

Candeias, M., "Going veg with *Nepenthes ampullaria*," In Defense of Plants blog, 2016.

Carroll, J., "What is a Sandbox tree: information About Sandbox tree exploding seeds," Gardening Know How, 2021.

Garden Fundamentals, www.gardenfundamentals.com

Gardenia, www.gardenia.net

Gardening with Angus, www.gardeningwithangus.com.au

Gilding, E.K. et al., "Neurotoxic peptides from the venom of the giant Australian stinging tree," *Science Advances*, 2020.

Hales, L., "Factsheet: Gympie-gympie," *Australian Geographic*, 2014.

"How one tree's venom inflicts months of agony," *Nature*, 2020.

Hurley, M., "'The worst kind of pain you can imagine'—what it's like to be stung by a stinging tree," The Conversation, 2018.

National Parks Flora & Fauna Web, www.nparks.gov.sg/florafaunaweb

Noushin Sharmili, S., "*Hura crepitans:* where nature produces 'explosives,'" Plantlet, 2021.

Pillai, S., "The suicide plant cause such excruciating pain it gives its victims suicidal thoughts," Sccop Whoop, 2016.

Plant World Seeds, www.plant-world-seeds.com

Plantopedia, www.plantopedia.com

Salt, A., "*Aldrovanda vesiculosa,* the Fly trap's endangered cousin," Botany One, 2018

Wan Zakaria, W. et al., "RNA-seq analysis of *Nepenthes ampullaria*," *Front, Plant Sci*, 2016.

Mistaken Identity

Bompas and Parr, www.bompasandparr.com

Byrd, D., "Brightest biological substance reveals its secret," Earth Sky, 2012.

Dyer, M.H., "*Aristolochia* pipevine plants: is growing Darth Vader flowers possible," Gardening Know How, 2021.

Dyer, M.H., "Flying Duck Orchid care—can you grow Flying duck orchid plants," Gardening Know How, 2021.

Gardens by the Bay, www.gardensbythebay.com.sg

The Garden of Eaden, www.gardenofeaden.blogspot.com

Grant, B.L., "Swaddled Babies Orchid: information about *Anguloa uniflora* care," Gardening Know How, 2021.

iNaturalist NZ, www.inaturalist.nz

The International Palm Society, www.palmtalk.org

Kuo, M. "*Xylaria polymorpha*," Mushroom Expert, 2019.

Raya Garden, www.rayagarden.com

Science and Plants for Schools, www.saps.org.uk

Strange Wonderful Things, www.strangewonderfulthings.com

Stromberg, J., "This African fruit produces the world's most intense natural color," *Smithsonian Magazine*, 2012.

Vignolini, S. et al., "Pointillist structural color in *Pollia* fruit," *PNAS*, 2012.

White, S., "*Aristolochia Salvadorensis*: The Darth Vader flower," Freaked blog, 2020.

Greater Good

American Peanut Council, www.peanutsusa.com

Florida Fruit Geek, www.floridafruitgeek.com

Freitas, B.M. and Paxton, R.J., "The role of wind and insects in cashew (*Anacardium occidentale*) pollination in NE Brazil," *The Journal of Agricultural Science*, 2009.

Gerald Mason, E., "What is a Cashew Apple and what does it taste like?", Mashed, 2021.

Hammons, R.O., et al. *Peanuts: Genetics, Processing, and Utilization*, AOCS Press, 2016.

Homegrounds, www.homegrounds.co/uk

Lausen-Higgins, J., "A taste for the exotic: pineapple cultivation in Britain," *Historic Gardens*, 2010.

Levins, H., "Social history of the pineapple," www.levins.com

Mongabay, www.mongabay.org

Superheroes

Adams, H., "*Wolffia globosa*: all about this super green supplement," *Clean Eating*, 2021.

Barras, C., "The secret of the world's largest seed revealed," *New Scientist*, 2015.

Bradford, A. and Dobrijeciv, D., "Corpse flower: facts about the smelly plant," Live Science, 2021.

The Gymnosperm Database, www.conifers.org

Kim, S.E., "Cultivating the world's largest stinkiest flower is no small task," *National Geographic*, 2021.

Korotkova, N. and Barthlott, W., "On the thermogenesis of the Titan arum (*Amorphophallus titanum*)," *Plant Signal Behaviour*, 2009.

Kurihara, K. and Beidler, L.M., "Taste-modifying protein from Miracle Fruit," *Science*, 1968.

Malaysian Species, www.malaysianspecies.weebly.com

NatureServe Explorer, https://.explorer.natureserve.org

Patowary, K., "Coco de mer: the forbidden fruit," Amusing Planet, 2017.

Plant Profiles in Chemical Ecology, www.sites.evergreen.edu/plantchemeco

Rafsanjani, A. et al., "Hydro-responsive curling of the Resurrection Plant *Selaginella lepidophylla*," *Scientific Reports*, 2015.

Sankey, S., "Miraculin: the miracle in Miracle Fruit," Botany One blog, 2019.

Sears, Cori. "How to grow Resurrection Plant (*Selaginella lepidophylla*), The Spruce, 2022.

WebMD, www.webmd.com

The Wollemi Pine, www.wollemipine.com

X-Rated

Bourebaba, L. et al., "Phytochemical composition of *Ecballium elaterium* extracts with antioxidant and anti-inflammatory activities: comparison among leaves, flowers and fruit extracts," *Arabian Journal of Chemistry*, 2020.

Feedipedia, www.feedipedia.org

Natural and Medicinal Herbs, www.naturalmedicinalherbs.net

New Forest National Park, www.newforestnpa.gov.uk

Oguis, G.K. et al., "Butterfly pea (*Clitoria ternatea*), a cyclotide-bearing plant with applications in agriculture and medicine," *Frontiers in Plant Science*, 2019.

South Africa National Biodiversity Institute, www.sanbi.org

Wisconsin Horticulture, https://hort.extension.wisc.edu

INDEX

Note: page numbers in **bold** refer to illustrations.

M

marble berry 66–9, **67–8**
Mayans 91
melon 116
Mesoamerican culture 104
Methuselah (Great Basin bristlecone pine) 139, 140
Mexican cotton *see* cotton
Microhyla nepenthicola (micro frog) 27
Mimosa pudica 7
miracle berry 8, 126–9, **126–7**
miracle fruit *see* miracle berry
miraculin 126–9
monarch butterfly 171
monkey no climb tree *see* dynamite tree
monkey nut *see* peanut
monkeys, capuchin 110–11
movement, plant 14, 119–21, 172–5
mucilage 30
mycelium 64
mycorrhizal fungi 179

N

narrow-lid Pitcher plant **24**, 25–7, **26**
National Parks and Wildlife Services (NPWS) 152–3
nectar 25, 35, 49, 108–10, 132, 170
Nepenthes ampullaria see narrow-lid pitcher plant
Netherlands 168
neurotoxins 16
New York Botanical Garden 136
Niigata botanical gardens 48
Niigata Prefectural Botanical Garden 120
nitrogen fixation 103, 167
Noble, David 152
noisy plants 40–3, **40–1**, **43**
NPWS *see* National Parks and Wildlife Services
nutcracker, Clark's 141

O

octopus stinkhorn **160–1**, 161–3, **163**
oldest trees 138–41
Olmec Indians 90–1
orchid
   cradle *see* orchid, tulip
   flying duck 9, 54–7, **55**, **57**, 159
   Italian *see* orchid, naked man
   monkey 47
   naked man 156–9, **157–8**
   tulip 74–7, **75**, **77**
Orchid Society of Great Britain 77
Orchis italica see orchid, naked man
overwatering 72–3

P

Palicourea elata see hot lips
parasites 146–9
Parish, Sarah 117
peanut 100–3, **100–2**
pepper 116
pericarps 174
petunia, climbing 114
Phallus impudicus see stinkhorn, common
photosynthesis 64, 72, 120, 124, 141
Physostegia virginiana 7
Phytophthora 152
pine, Great Basin bristlecone 138–41, **138–40**
pineapple plant 96–9, **96–7**, 99
pinnate foliage 23
Pinus longaeva see Great Basin bristlecone pine
Pitcher plant, narrow-lid **24**, 25–7, **26**
poinsettia 60
pollen 29, 49, 52, 56–7, 76, 130, 132, 170
Pollia condensata see marble berry
pollination 29, 42
   Asian watermeal 142, 144
   bullhorn acacia 32
   burning bush 20

AUTHOR'S ACKNOWLEDGMENTS

This book is dedicated to anyone who has ever been intrigued by plants, but is put off by all those big, heavy, and serious books, with wall-to-wall Latin names and unfathomable botanical terms. We don't always need to be botanists to share a passion for plants, and to entice others onto the journey.

I'd like to start off with a huge thanks to DK for indulging my crazy love of plants, and helping to realize it into an all-singing all-dancing book. Thanks to Katie Cowan, Max Pedliham, Chris Young, Ruth O'Rourke, Louise Brigenshaw, Abi Read, and Izzy Holton. And, of course, this book would also be pretty dry without its gorgeous illustrations—a massive thanks to Aaron Apsley and his talented hand.

With my limited technical botany experience, I am forever indebted in particular to Alastair Robinson and Marc Hachadourian. I was humbled to have them both on botanical speed-dial, albeit at opposite ends of the planet. Plants will always bring people together, something we are discovering more and more. Thank you also to a bevy of other botanical buddies—there are too many to name—but I have hugely appreciated your input in the making of this book, and the many ways you all inspire me. And, in the years preceding, thanks to those who gave me privileged insights into some of the world's best botanical gardens, from London to Leiden, and, of course, the wild adventures, from Sarawak to the Seychelles.

The biggest of thanks though, to my now-departed grandparents, who ignited this spark in me at the tender age of 5 years old. Thanks to my college friends who allowed it to blossom (but not to the school bullies who didn't). Thanks also to my long-suffering partner and friends, who have allowed me to catfish them into all manner of botanical visits over the years.

I hope you have enjoyed this book, whether you're a plant obsessive, complete novice, or anything in between.

AUTHOR BIO

Michael Perry, aka **Mr. Plant Geek**, is a leading gardening personality. Formerly the new plant developer for established seed and plant company, Thompson and Morgan, he is now a TV host, writer, and broadcaster. He presents unique slots about plants and growing on ITV, Channel 4, and QVC in the UK, and is a well-known social media presence. He won the Garden Media Guild Social Media Influencer of the Year Award in 2020, and the GMG's Broadcast Award in 2021. His presence extends to Europe, the US, Japan, and Canada, where he has traveled extensively, picking up new ideas and seeing many of his dream plants in real life.

ILLUSTRATOR BIO

Aaron Apsley is a botanical illustrator based in Florida. He spent his childhood exploring the woods in rural Appalachia, and cultivated a love for the natural world at an early age. After studying illustration and painting in college, Aaron moved to New York City, where he began growing and painting houseplants as a way to reconnect with nature in an urban environment. In 2015, he began sharing his illustrations on the instagram account @apsley_watercolor, where he uses botanical art to highlight the beauty and diversity that can be found in the world of plants. Moving from New York to the subtropical climate of southern Florida in 2019 opened up a whole new world of botanical inspiration for Aaron, and several of the watercolor illustrations in this book are based on plants that he is growing in his own garden. When he isn't painting, Aaron is tending to his collection of tropical plants, traveling to parks and botanical gardens, and connecting with people through his love of botany.

Picture credit: Copyright © Snooty Fox Images